Laser Speckle
and Applications in Optics

Laser Speckle
and
Applications in Optics

M. FRANÇON

Optics Laboratory
Faculty of Sciences
University of Paris
Paris, France

Translated by HENRI H. ARSENAULT

Department of Physics
Laval University
Quebec, Canada

ACADEMIC PRESS   New York   San Francisco   London   1979

A Subsidiary of Harcourt Brace Jovanovich, Publishers

√63106711

PHYSICS

ACADEMIC PRESS, INC.
111 Fifth Avenue, New York, New York 10003

United Kingdom Edition published by
ACADEMIC PRESS, INC. (LONDON) LTD.
24/28 Oval Road, London NW1   7DX

Library of Congress Cataloging in Publication Data

Françon , Maurice.
    Laser speckle and applications in optics.

    Bibliography: p.
    1. Laser speckle. 2. Optics. I. Title.
TA1677.F73   621.36'6   79—13043
ISBN 0—12—265760—8

French edition published as *La Granularite Laser (Speckle)*
et ses Applications en Optique.
© Masson, Editeur, Paris, 1977.

PRINTED IN THE UNITED STATES OF AMERICA

79 80 81 82   9 8 7 6 5 4 3 2 1

Contents

Preface   xi

Chapter I   **Speckle in the Image of an Object Illuminated
             with Laser Light**

1.1   Image of a Point Source. Fourier Transforms   1
1.2   The Image of a Point Source in the Presence of a Slight Defect
       of Focus   4
1.3   The Images of Two Monochromatic Point Sources   5
1.4   The Images of a Large Number of Point Sources Distributed
       at Random   8
1.5   The Spectrum of a Large Number of Coherent Point Sources   9
1.6   The Spectrum of a Large Number of Coherent Point Sources Forming
       Identical Groups Having the Same Orientation and Distributed at
       Random   11
1.7   Speckle in the Image of an Object Illuminated with a Laser   14
1.8   Changing the Structure of a Speckle by Displacing the Plane of Focus   16
1.9   The Speckle Patterns Produced in the Image of a Diffuse Object
       under a Change of Wavelength   17
1.10   White-Light Speckle   19

Chapter II   **Speckle Produced at a Finite Distance by
              a Diffusing Object Illuminated by a Laser**

2.1   Fresnel and Fraunhofer Diffraction in Three Dimensions   21
2.2   Speckle Produced at a Finite Distance by a Diffuse Object   24
2.3   Speckle Produced by a Laterally Displaced Diffusing Object   25
2.4   Speckle Produced by a Diffuse Object When the Orientation of the
       Incident Light Beam Is Changed   26
2.5   Speckle Produced by a Diffuse Object under an Axial Translation
       of the Plane of Observation or of the Object   28
2.6   Speckle Produced by a Diffuse Object under a Change of
       Wavelength   30
2.7   Speckle Produced by a Diffuse Object When the Wavelength and the
       Position of the Plane of Observation Are Changed   31
2.8   Speckle Produced by a Diffuse Object Itself Illuminated with Another
       Speckle Pattern   33

Chapter III   **Interference with Scattered Light**

3.1   Historical Background   35
3.2   Principles of Interference with Scattered Light   36
3.3   Interference with Two Identical Diffusers   41
3.4   The Burch Interferometer   42
3.5   Interference Patterns Obtained from Two Laterally Shifted Diffusers   43
3.6   Interference Patterns Obtained from Two Different Axially
       Shifted Diffusers   47

Chapter IV   **Interference Patterns Produced by Photographic
              Superposition of Laterally Shifted Speckle
              Patterns**

4.1   Amplitude Transmitted by a Photographic Plate after Development   49
4.2   The Fundamental Experiment of Burch and Tokarski   50
4.3   The Superposition of a Series of Successive Exposures on the Same
       Holographic Plate   53
4.4   Simultaneous Recordings with Birefringent Plates   55
4.5   Recording with a Continuous Displacement of the Photographic
       Plate   58
4.6   Recordings Obtained When the Orientation of the Incident Beam
       Is Changed   59
4.7   Recordings Made with Polarized Light   60
4.8   Recordings Obtained When the Orientation of the Diffuse Object
       Is Changed   61
4.9   Recordings Made with More Than One Wavelength   62

Chapter V   **Interference Patterns Produced by Photographic Superposition of Axially Shifted Speckle Patterns**

5.1   Circular Interference Fringes Produced by Two Successive Recordings on the Same Photographic Plate   65
5.2   A Change of Wavelength between the Two Exposures   67
5.3   A Change of Wavelength and a Displacement of the Photographic Plate between Exposures   68
5.4   Circular Interference Fringes Produced by Two Successive Recordings on the Same Photographic Plate of the Image of a Diffuse Object   69
5.5   Circular Interference Fringes Obtained with a Single Exposure by Means of an Amplitude Diffuser   69
5.6   Circular Interference Fringes Obtained with More Than One Exposure When the Photographic Plate Is Axially Translated between Each Exposure   70
5.7   Hyperbolic or Elliptical Fringes   72

Chapter VI   **Optical Processing of Images Modulated by Speckle**

6.1   Introduction   73
6.2   The Principle of a Technique to Extract the Difference between Two Images   75
6.3   The Light Distribution in the Image Plane   77
6.4   Improving the Profile of the Fringes and the Quality of the Images   79
6.5   Image Coding and Decoding   83
6.6   Image Multiplexing by Superposition of Laterally Shifted Speckle Patterns   84
6.7   Image Multiplexing with Oriented Speckle Patterns   85

Chapter VII   **The Study of Displacements and Deformations of Diffuse Objects by Means of Speckle Photography**

7.1   The Study of Lateral Displacements of a Diffuse Object When the Displacement Is Greater Than the Diameter of a Speckle Grain   89
7.2   The Study of Lateral Displacements of a Diffuse Object When the Displacement Is Smaller Than the Diameter of a Speckle Grain   92
7.3   The Study of Lateral Displacements of a Diffuse Object Illuminated with Two Beams When the Displacement Is Smaller Than the Diameter of a Speckle Grain   93
7.4   The Study of the Lateral Displacement of a Diffuse Object with a Diffuse Reference Surface   95

7.5   The Study of the Axial Displacement of a Diffuse Object with a
      Diffuse Reference Surface   96
7.6   The Speckle Observed in the Focal Plane of Lens *O*   98
7.7   The Use of an Auxiliary Speckle Pattern to Illuminate a
      Diffuse Surface   99
7.8   The Study of the Rotation of a Diffuse Surface   100
7.9   The Study of the Vibrations of a Diffuse Object   102
7.10 The Study of the Variations of the Slopes of a Diffuse Object   106

Chapter VIII   Speckle in Astronomy

8.1   The Image of a Star at the Focus of a Telescope in the Presence of
      Atmospheric Turbulence   111
8.2   The Study of Double Stars at the Focus of a Telescope in the
      Presence of Atmospheric Turbulence   113
8.3   The Measurement of the Apparent Diameters of Stars by
      A. Labeyrie's Methods   115
8.4   The Measurement of the Apparent Diameters of Stars with
      Multiple Telescopes   118

Chapter IX   The Study of Surface Roughness

9.1   Surface Deviations   121
9.2   The Use of Speckle to Study Surface Roughness   122
9.3   Surface Roughness Measurement by the Correlation of Two
      Speckle Patterns Obtained by Changing the Incidence of a
      Laser Beam   122
9.4   The Real-Time Measurement of Surface Roughness by the Amplitude
      Correlation of Two Speckle Patterns Corresponding to Two
      Orientations of the Laser Beam   124
9.5   Surface Roughness Measurement by the Correlation of Two
      Speckle Patterns Obtained with Two Wavelengths   127
9.6   Surface Roughness Measurements with a Source Having a Wide
      Bandwidth   128
9.7   Surface Roughness Measurements with Partially Coherent Light   129

Chapter X   Various Applications of Speckle

10.1 Differential Interferometry of Transparent Objects by
      Double-Exposure Speckle Photography   133
10.2 The Use of Two Wavelengths to Study the Shape of Diffuse Objects   134
10.3 The Use of Speckle to Determine the Transfer Function of an
      Optical System   136
10.4 The Study of the Aberrations of an Optical System by Speckle
      Photography   137

10.5	The Focusing of a Lens by Means of Speckle	138
10.6	Laser Speckle for Determining the Ametropia of the Eye	140
10.7	The Use of the Double Exposure of a Random Distribution of Intensity to Measure Atmospheric Turbulence	141
10.8	The Measurement of Motion Trajectories by Speckle Photography	142
10.9	The Determination of the Velocities of Different Parts of a Diffuse Object by Speckle Photography	144
10.10	An Example of Industrial Application of Speckle Interferometry	146

References 149

Index 159

Preface

When an object is illuminated by a laser, it appears to be covered with a very fine granular structure. For this it is necessary that the object diffuse the light, as would occur, for instance, with a piece of paper, a cement surface, or an unpolished metal surface. All the points of the object illuminated by the laser are coherent, and the waves that they transmit to the eye are capable of interfering. The image from each object point produces on the retina a diffraction image that depends upon the imaging system of the eye. It is the interference between these diffraction images that is responsible for the granular structure known as speckle. If the eye is replaced by a photographic system, the phenomenon is the same: after development, the image displays a speckle pattern that depends upon the aperture of the lens. When the aperture is wider, the speckle is smaller. This is to be expected, because the size of the diffraction pattern of a lens decreases as the aperture of the lens is increased. However, the formation of an image is not necessary to obtain speckle. A diffuse object illuminated by a laser produces a speckle pattern in the space surrounding it. A photographic plate placed at any distance from the object may record the speckle pattern. By analogy with diffraction phenomena, speckle may be said to be of the Fraunhofer type in the first case, and of the Fresnel type in the second.

When the hologram of an object is recorded, speckle appears in the reconstructed images, causing a degradation of image quality. For this reason, considerable effort has been put into speckle reduction techniques. On the other hand, it was not long before speckle led to some very useful applications, which are the subject of this monograph. The development of speckle techniques in the past few years has been such that there is little doubt that a new chapter in optics has been opened.

In this work on speckle optics, the emphasis is on the experimental aspect of phenomena and on applications, but we have retained the theoretical basis founded on Fourier optics.

In the first chapter the elements of diffraction theory are reviewed; we then study the properties of speckle in the image of a diffuse object. Chapter II progresses along similar lines, but here we consider speckle in the near field. Most of the experiments carried out with speckle are experiments of interference that may be produced with any diffusers. Chapter III is therefore dedicated to interferometry with diffuse light. In Chapter IV the interference patterns produced from photographically superimposed laterally shifted speckle patterns are considered. These experiments consist essentially of superimposing two or more speckle patterns on the same photograph, the emulsion being shifted slightly after each exposure. The starting point of all the present techniques using speckle is no doubt the fundamental experiment of Burch and Tokarski. In Chapter V we again consider similar experiments, where the translation this time is in a direction perpendicular to the plane of detection.

Some applications of those experiments are given in the five chapters that follow. In Chapter VI we consider optical processing of images modulated by speckle, whereas in Chapter VII we study the deformations and displacements of diffuse objects. Chapter VIII is dedicated to applications in astronomy, and Chapter IX to surface roughness measurements. Finally, in Chapter X we consider various applications: the study of transparent objects, the average shape of diffuse surfaces, the transfer functions and aberrations of optical systems, and the movement of diffuse objects.

A word on astronomy: by considering the turbulence of the atmosphere to be equivalent to a diffuser, Antoine Labeyrie has carried out one of the most beautiful experiments in optics. His remarkable method, which has allowed him to study double stars and to measure the apparent diameter of stars, opens up some extraordinary possibilities.

The number of authors who have published research on speckle techniques is quite considerable, and it has not been possible to mention them all throughout the chapters of this book. We hope that they will forgive us.

The references given in the sections are not a complete bibliography of the questions to which they refer. They are meant only as a guide to the reader, who will find at the end of this monograph a more detailed bibliography.

CHAPTER I

Speckle in the Image of an Object Illuminated with Laser Light

1.1   Image of a Point Source. Fourier Transforms

Let an objective lens O, assumed to be perfect, be illuminated by a point source S emitting monochromatic light of wavelength λ (Fig. 1). The spherical wave Σ from S is transferred by the lens into a spherical wave Σ' whose center S' is the geometrical image of source S. It is well known that the real image at S' is a small spot of light whose structure is determined by diffraction phenomena. This small spot, called the diffraction image, depends upon the shape of the periphery of lens O. To study the structure of this diffraction image is to study diffraction at infinity, or the Fraunhofer diffraction pattern. We say diffraction at infinity because the lens O may be replaced by two lenses whose focal lengths are twice as great. The source S is then at the front focal plane of the first lens, and the image S' is at the back focal plane of the second lens. The second lens is thus illuminated by a source at infinity.

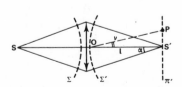

Fig. 1   Image S' of a point source S.

From the shape of the emergent wave Σ', the structure of the diffraction image of the point source S may be calculated by using the Huygens–Fresnel principle. Assume that the angular aperture 2α in the image space is not too great, so that $\cos \alpha$ is approximately equal to one. It may be shown that the Huygens–Fresnel principle allows the diffraction phenomena to be expressed by means of the Fourier transform. The amplitude at any point P in the plane π' is given by the Fourier transform (or spectrum) of the complex amplitude on the surface of the wave Σ'. Inversely, the complex amplitude on the surface of Σ' may be calculated if the amplitude and phase in the diffraction image S' are known. The complex amplitude on Σ' is the inverse Fourier transform of the complex amplitude in the diffraction image. The use of the Fourier transform and its properties allows the calculation of diffraction phenomena in a simple and elegant way.

Consider, for example, a lens O with a circular aperture. The amplitude in the image plane π' is given by the Fourier transform of a circular surface. Let $2a$ be the diameter of the lens O, and let P be any point in the plane determined by the angle v. Let

$$Z = Kav, \qquad K = \frac{2\pi}{\lambda} \tag{1.1}$$

The amplitude at P is equal to

$$f(v) = \frac{2J_1(Z)}{Z} \tag{1.2}$$

This expression is a real function, where $J_1(Z)$ is the Bessel function of order one of the variable Z. Figure 2 shows the variations of amplitude $f(v)$, and Fig. 3 the variations of intensity $I = |f(v)|^2$ as a function of Z. Of course the diffraction image or Airy disk has symmetry of revolution. It consists of a very luminous central spot, surrounded by alternating dark and bright rings. Every other ring is seen to have a negative amplitude. The intensities of the luminous rings are much smaller than that of the central spot, and

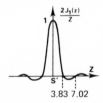

Fig. 2   Amplitude diffraction pattern of a circular aperture.

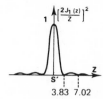

Fig. 3   Intensity diffraction pattern of a circular aperture.

decrease rapidly. For the first dark ring, the table of Bessel functions gives $Z = 3.83$. From (1.1), the angular radius v of the first dark ring is equal to

$$v = \frac{1.22\lambda}{2a} \tag{1.3}$$

The diameter of the central diffraction spot, equal to $2v$, is also known as the angular diameter of the diffraction image. It increases inversely with the diameter $2a$ of the lens O. If $l = os'$ and setting $\alpha = a/l$, the linear diameter of the diffraction image in plane π' (Fig. 1) is equal to

$$2vl = \frac{1.22\lambda l}{a} = \frac{1.22\lambda}{\alpha} \tag{1.4}$$

For example, a lens with aperture 2α and a wavelength $\lambda = 0.6 \ \mu m$ (yellow) will yield a diffraction image with a diameter equal to 3 μm.

Figure 4 shows the diffraction phenomena for three types of apertures. Figure 4a shows the diffraction image when a screen with a slit of width ζ_0 is put on the lens. The curve of Fig. 4b represents the diffraction image produced when a screen with a narrow angular aperture is used with the lens. Finally, Fig. 4b represents the diffraction image when the circular lens

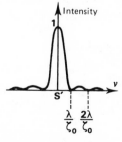

Fig. 4a   Diffraction pattern of a slit of width ζ_0.

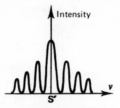

Fig. 4b   Diffraction pattern of an annular aperture.

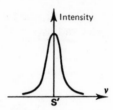

Fig. 4c   Diffraction pattern of a circular aperture when the absorption decreases from the center to the edge.

does not have a uniform transparency. If the absorption increases from the center to the edge according to a Gaussian distribution, the intensity in the diffraction image is also a Gaussian distribution.

1.2   The Image of a Point Source in the Presence of a Slight Defect of Focus

Let us displace the plane of detection from π' to π'' (Fig. 5), and observe the image of the point source S on plane π'', whose distance from plane π' is equal to δl. We assume that the displacement δl is small compared to distance OS''. The geometrical image S'' of S is the center of a spherical wave Σ, assuming a perfect lens. The vibrations from different points of the surface wave Σ are in phase at S', but not at S''. The maximum path difference at S''

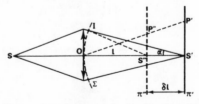

Fig. 5   Image of a point source in the presence of defocus.

is equal to $\Delta = IS'' - OS''$, and is responsible for the degradation of the image at S''.

It is easy to show that

$$\Delta = \delta l \frac{\alpha^2}{2} \tag{1.5}$$

If it is required that the diffraction image at S'' be practically identical to that at S', then Δ must be much smaller than the wavelength λ of the light emitted by S. This leads to the condition

$$\delta l \ll \frac{2\lambda}{\alpha^2} \tag{1.6}$$

In order to understand the diffraction phenomena, it is useful to study the distribution of light intensity in the image S'. Figure 6 shows the isophotes in the region about S'. The greatest energy density is found in a cigar-shaped volume whose length is equal to $4\lambda/\alpha^2$, and whose maximum width is equal to $1.22\lambda/\alpha$, the diameter of the diffraction image. The cigar shape is the central part of Fig. 6.

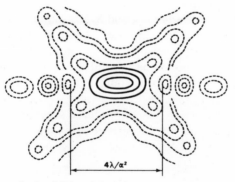

Fig. 6   Isophotes in the vicinity of the image of a point source (circular aperture).

1.3   The Images of Two Monochromatic Point Sources

We now consider two point light sources S_1 and S_2, obtained as shown in Fig. 7. Two small holes S_1 and S_2 are made in an opaque screen π. They are illuminated with an extended uniform monochromatic light source S_0. The two small holes S_1 and S_2 diffract the light from S_0, and behave practically like a pair of point sources. The two images S_1' and S_2' formed by the lens O are diffraction images identical to the diffraction image studied previously (Fig. 2). If S_1 and S_2 are illuminated only by a small element dS_0 about point M in source S_0, the light waves from S_1 and S_2 have a path difference Δ equal to $MS_1 - MS_2$, and a phase difference $\varphi = 2\pi\Delta/\lambda$, where λ is the wavelength of the light. The two sources S_1 and S_2, when illuminated by the small source dS_0, are coherent.

To study the phenomena in plane π', the amplitudes must be added, taking into account the phase differences φ. For example, if $\varphi = \pi$, the amplitudes of the two diffraction images must be subtracted, as shown in Fig. 8. If S_1 and S_2 were illuminated by another element of source S_0, the two diffraction images S_1' and S_2' would not change, but their phase difference would not be the same, and the intensity distribution in plane π' would be changed. Figure 9 illustrates the case when the two diffraction images are in phase.

Now let S_1 and S_2 be illuminated by the two previous elements of the source; the intensity distributions of Figs. 8c and 9c must now be added, because the elements of source S are in effect incoherent, that is, there is no phase relationship between them, and in such cases it is the intensities of the diffraction images that must be added.

Consider finally the case where S_1 and S_2 are illuminated by the whole of source S_0. The preceding operation is repeated for all the elements of source S_0, and it is found that the intensity distribution in plane π' is equal to that obtained when the intensities of the two diffraction images are added: when S_1 and S_2 are illuminatrd by an extended source S_0, they behave like two incoherent sources, and the intensity in plane π' is found by adding the intensities of the two diffraction images produced separately by S_1 and S_2.

The foregoing results may be summarized by using the Dirac delta function: if the sources S_1 and S_2 are coherent, the amplitude distribution in

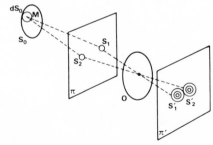

Fig. 7   Images (diffraction patterns) S_1' and S_2' of two apertures S_1 and S_2.

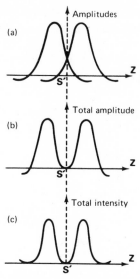

Fig. 8   (a) Amplitude diffraction patterns. (b) Total amplitude when the amplitudes are in opposition. (c) Total intensity.

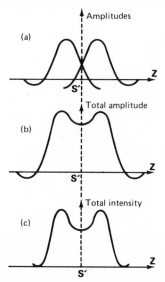

Fig. 9   (a) Amplitude diffraction patterns. (b) Total amplitude when the amplitudes are in phase. (c) Total intensity.

plane π' is given by the convolution of the amplitude of the diffraction image of the lens O with two delta functions corresponding to the two geometrical images S_1' and S_2'. If the sources S_1 and S_2 are incoherent, the intensity distribution in the plane is given by the convolution of the intensity of the diffraction image of the lens O with the two delta functions.

1.4   The Images of a Large Number of Point Sources Distributed at Random

Going back to the setup shown in Fig. 7, let the screen π now have a distribution of identical small holes S_1, S_2, S_3, etc., distributed at random as shown in Fig. 10. Consider first the case where S_1, S_2, S_3, etc., are illuminated only by a small element dS_0 of the source S_0, about a point M. The small apertures S_1, S_2, S_3, etc., diffract the incident light and act as coherent light sources. The image of each source in plane π' is a diffraction image of the type shown in Fig. 2. All the diffraction images produced by S_1, S_2, S_3, etc., are identical, and in order to study the phenomena in plane π', the amplitudes must be summed as in the previous case, while taking into account the phase relationships. The diffraction images more or less overlap, and a quite complicated structure is obtained, consisting of small spots, the smallest of which have diameters approximately equal to that of the diffraction spot of lens O when it is illuminated by a point source (1.3). This granular structure is called "speckle." Figure 11 shows what happens when we have 11 identical diffraction images distributed at random on a straight line. Figure 11a shows the amplitudes of the diffraction images; the phases are identified by the parameter φ in degrees. Figure 11b shows the resulting intensity distribution.

If the size of element dS_0, which illuminates S_1, S_2, S_3, etc., is now increased, the latter now behave like partially coherent sources. The contrast of the speckle decreases. If the apertures S_1, S_2, S_3, etc., are illuminated by the extended source S_0, the contrast finally tends to zero and the speckle

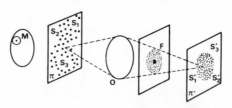

Fig. 10   Image given by a lens O from a screen π with a large number of small randomly distributed apertures.

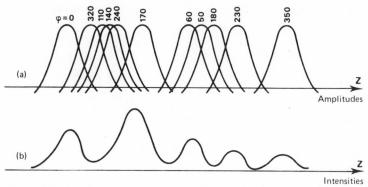

Fig. 11   Amplitude diffraction patterns with different phases φ (upper curves), and the resultant intensity (lower curve).

disappears. In this case, sources S_1, S_2, S_3, etc., are incoherent, and if the sources are close enough together, the plane has a uniform illumination.

Using the delta distribution, the speckle may be said to result from the convolution of the amplitude of the diffraction image of the lens O with the set of delta functions corresponding to the geometrical images S_1', S_2', S_3', etc., corresponding to S_1, S_2, S_3, etc. If the sources S_1, S_2, S_3, etc., are incoherent, the convolution is with the intensity of the diffraction image.

1.5   The Spectrum of a Large Number of Coherent Point Sources

For simplicity, we assume that plane π has many small apertures S_1, S_2, S_3, etc., and is illuminated by a point source at infinity (Fig. 12). The small apertures S_1, S_2, S_3, etc., behave like coherent sources, and we shall study what goes on in the focal plane F of the lens O. We observe in F the diffraction image, or spectrum, of the set of small apertures S_1, S_2, S_3, etc.,

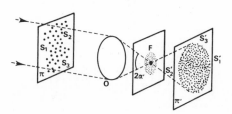

Fig. 12   Spectrum in F of a screen π with a large number of randomly distributed small apertures.

which are assumed to be all similar. Care must be taken to distinguish this case from the preceding one: in Section 1.3, we studied the images of S_1, S_2, S_3,... in plane π', and these images were the diffraction images of the lens O, because we had assumed point sources. On the other hand, in the focal plane of lens O, we observe the diffraction from the set of point sources S_1, S_2, S_3,.... This is also Fraunhofer diffraction, because the observation is in the focal plane of lens O. The diffraction phenomenon is the Fourier transform or spectrum of the set of point sources. All the waves emitted by the sources S_1, S_2, S_3,... are in phase at F, the back focal plane of lens O. The intensity is therefore maximum at this point. As we move away from F, the intensity decreases very rapidly. At an arbitrary point P, the phases of the waves sent out by S_1, S_2, S_3, etc., may take any values between 0 and 2π. At another nearby point, the phases of the waves sent out by S_1, S_2, S_3, etc., change, and even if the phase variations are small, they may cause important changes in the amplitude and in the intensity. These intensity fluctuations are responsible for the granular aspect of the phenomena observed in plane π'. In the focal plane F, there is also speckle whose origin is different from the speckle in plane π'. As before, the speckle in F is composed of small luminous spots, the smallest of which have diameters approximately equal to the diameter of the diffraction image of the lens O, in the focal plane F. In plane π', it is the diffraction image of the lens O, which has an aperture 2α corresponding to the distance from O to π, that counts. On the other hand, in the focal plane F the diffraction image that counts is that of the lens with an aperture $2\alpha'$ (Fig. 12) corresponding to the distance from O to F. Because $\alpha' > \alpha$, the speckle in F will have a finer structure than the speckle in π' (1.4).

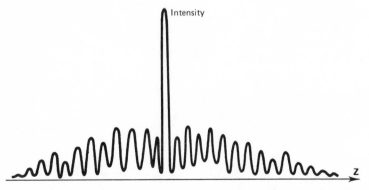

Fig. 13   Intensity distribution in the Fourier spectrum of a large number of small apertures distributed at random.

In practice, the sources S_1, S_2, S_3, etc., always have finite dimensions, and the spectrum does not spread out indefinitely in plane F. This is a classical problem, the results of which we shall briefly recall. If the apertures S_1, S_2, S_3, etc., are identical, oriented in the same way, and distributed in an irregular way in plane π, the diffracted light in plane F is modulated by the diffraction image of one of the apertures. This is illustrated in Fig. 13. A very intense diffraction image, which is practically the diffraction image of the lens O illuminated only by a point source at infinity, is found at the center. With N apertures, the intensity at the center is proportional to N^2, because all the vibrations are in phase at F. The diffracted light, which is the diffraction image of the apertures, is spread out about this bright spot. As we have said, this diffraction image is speckle.

The smaller the diameters of the apertures S_1, S_2, S_3, etc., the more the speckle spreads out in plane F (1.4). Plane π with its apertures S_1, S_2, S_3, etc., may be considered as a diffuser. Its spectrum is another diffuser that spreads out more in plane F as the diffuser in π has a finer structure.

1.6   The Spectrum of a Large Number of Coherent Point Sources Forming Identical Groups Having the Same Orientation and Distributed at Random

Let screen π contain groups of two small apertures such as $(S_{1.1}, S_{1.2})$, $(S_{2.1}, S_{2.2})$, etc. (Fig. 14). All the small apertures are assumed to be identical. The distance between two sources S_{i1}, S_{i2} of a group is the same for all groups, and is equal to ζ_0. The straight line joining two sources S_{i1}, S_{i2} of any group is in the same direction for any group. The groups are distributed at random. We may also consider the screen π to contain a great number of small apertures $S_{1.1}, S_{2.1}, S_{3.1}$, etc. The set of apertures $S_{1.2}, S_{2.2}, S_{3.2}$ may be obtained from this set by means of a simple translation equal to ζ_0. If the central region of F is neglected, it may be stated, as before, that the

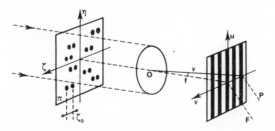

Fig. 14   Spectrum of a large number of groups of two apertures distributed at random.

spectrum of this set of apertures is the same as that of a group of two small apertures, but it is N times as luminous if there are N groups. The spectrum of a group of two small apertures consists of Young's fringes oriented in a direction perpendicular to a line joining two small apertures. The angular distance v between two consecutive bright fringes or between two consecutive dark fringes is

$$v = \frac{\lambda}{\zeta_0} \qquad (1.7)$$

Let the function $D(\eta, \zeta)$ represent the point sources $S_{1.1}, S_{2.1}, S_{3.1}, \ldots$ having the coordinates η, ζ. The screen π with all the apertures is represented by the expression

$$D(\eta, \zeta) \otimes [\delta(\eta, \zeta) + \delta(\eta, \zeta - \zeta_0)] \qquad (1.8)$$

where $\zeta(\eta, \zeta)$ is the **Dirac** delta function centered on η, ζ. A translation is in effect equivalent to convolution with a delta function, and on the other hand

$$D(\eta, \zeta) \otimes \delta(\eta, \zeta) = D(\eta, \zeta) \qquad (1.9)$$

The spectrum of the set of apertures in screen π is given by the Fourier transform of expression (1.7), where $D(\eta, \zeta)$ is the amplitude at a point having the coordinates η, ζ. The function $D(\eta, \zeta)$ represents a diffuser whose Fourier transform $\tilde{D}(u, v)$ is a speckle distribution in the focal plane Fuv. Now the Fourier transform of two delta functions is equal to $1 + \exp(j\pi v \zeta_0 / \lambda)$, where $j = \sqrt{-1}$; so except for the central region near F, the Fourier transform of the amplitude distribution (1.6) of screen π is equal to

$$\tilde{D}(u, v)[1 + \exp(j\pi v \zeta_0 / \lambda)] \qquad (1.10)$$

The intensity, except for a constant factor, is

$$I = |\tilde{D}(u, v)|^2 \cos^2\left(\frac{\pi v \zeta_0}{2\lambda}\right) \qquad (1.11)$$

The diffuse background $|D(u, v)|^2$ is modulated by a set of Young's fringes whose spacing is given by Eq. (1.6). The autocorrelation function of the screen π with all the apertures may also be used. It is represented by three signals of relative amplitude 1, 2, 1, and its Fourier transform, a function that varies as $\cos^2 x$, yields directly the intensity distribution in plane Fuv.

Consider now a set of small apertures oriented in the same direction. They behave like small luminous parallel slits distributed at random (Fig. 15). Suppose that all the slits have the same length ζ_0. All the slits are coherent.

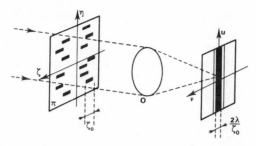

Fig. 15   Spectrum of a large number of small slits distributed at random.

If N is the number of slits, the spectrum at F is the same as that of a single slit, but N times as bright. Again we neglect the intense narrow peak located at F. In plane Fuv may be observed the spectrum of a slit whose intensity variations are oriented in a direction Fv parallel to the slits. The slits are assumed to be infinitely thin, and there are therefore no intensity variations in a direction perpendicular to the slits. Not taking into account the diffraction fringes, the diffracted light in the spectrum may be said to be concentrated in a narrow band of width $2v = 2\lambda/\zeta_0$.

Consider the set of point sources formed by one of the ends of each small slit. This set may be represented by a function $D(\eta, \zeta)$ as before. The screen π with all the small slit apertures is then represented by

$$D(\eta, \zeta) \otimes \text{Rect}\left(\frac{\zeta}{\zeta_0}\right) \tag{1.12}$$

where each slit is obtained from a convolution of one of the end points with a rectangle function of length ζ_0. The spectrum in plane Fuv is given by the expression

$$\tilde{D}(u, v) \frac{\sin(\pi v \zeta_0/\lambda)}{\pi v \zeta_0/\lambda} \tag{1.13}$$

And the intensity is equal to

$$|\tilde{D}(u, v)|^2 \left(\frac{\sin(\pi v \zeta_0/\lambda)}{\pi v \zeta_0/\lambda}\right)^2 \tag{1.14}$$

The diffuse background $|\tilde{D}(u, v)|^2$ is modulated by the function $\text{sinc}(\pi v \zeta_0/\lambda)$, the diffraction image of a slit of width ζ_0. This result could also be obtained from the Fourier transform of the autocorrelation function of the rectangle function, which would directly yield the intensity.

1.7   Speckle in the Image of an Object Illuminated with a Laser

Diffuse objects are usually of two types: diffuse reflecting objects and transparent diffuse objects. Variations of thickness, of reflection, of absorption, and changes in the index of refraction are the causes that generally produce the scattering of light. In the case of objects commonly known as "diffuse objects," these variations are great compared to the wavelength of light. The rough surfaces of stones, cement, unpolished metal, wood, and so on are diffuse reflecting objects. Ground glass is a well-known example of a transparent diffuse object. It is difficult to determine a lower limit under which an object is considered well polished, because it is not only the thickness variations that count, but their spatial distribution: in other terms, the slopes of the profile of the diffuse object. These are no more than a minute for ordinary glass, and less than a second of arc for optically polished surfaces. It goes without saying that the foregoing figures are only orders of magnitude.

Consider any diffuse object, for instance, a transparent diffuse object such as a ground glass. It is illuminated by an extended light source. Form an image of this object with a lens. Each point of the object has a diffraction pattern whose dimensions depend only on the image-forming lens used under the given conditions. All the points of the object illuminated with the light source are incoherent, and the image of the ground-glass diffuser yields a superposition in intensity of all the diffraction patterns due to the different points of the ground glass. Because the object is uniformly illuminated, so is the image. This is the case studied in Section 1.4.

When a diffuse object G is illuminated with a laser as in Fig. 16, that is, by what is effectively a point source emitting spatially and temporally coherent light, all the points of the object scatter coherent light waves that are capable of interfering. When the object is looked at, a very fine random structure of small bright points is seen in the image: this structure, known as "speckle," is caused by interference in the image between the waves from

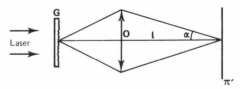

Fig. 16   Speckle in the image π' of a ground glass G illuminated with a laser (transparent diffuse object).

the various points of the object. This is the case studied in Section 1.5. The photograph shown in Fig. 86 shows the structure of speckle magnified many times with a microscope.

The image is the result of the superposition of the amplitudes of the diffraction patterns due to the different points of the object, taking into account the phase changes produced by the thickness variations of the diffuser. A large number of small bright spots distributed at random are seen; the smallest spots have diameters that are of the order of magnitude of the diameter of the diffraction pattern of the image-forming lens. Let the diameter of the lens O be equal to $2a$ (Fig. 16) and the distance between the lens O and the plane of observation π' be equal to l. If $\alpha = a/l$, then from Eq. (1.5)

$$\varepsilon \simeq \frac{\lambda}{\alpha} \tag{1.15}$$

Because the object G is illuminated with a parallel beam of light with an arbitrary angle of incidence, any lateral translation of the object in its mean plane does not change the relative phases of the various object points, and in the image plane π', the speckle does not change, but follows the object in its translation. This is an important point that will be used later on. If the object is rotated in its plane, the relative phases are modified, and so is the speckle, except if the incident light is perpendicular to the plane of the object. The phenomena are the same for a diffuse reflecting object. Such an object, whose thickness variations oscillate about a mean plane, is illuminated with an inclined laser beam as shown in Fig. 17. Any lateral translation of the object in its mean plane will keep the relative phases from the various elements of the object invariant, and in the image plane π' the speckle is not modified. If a rotation of the object is not to change the speckle, then it is required to use normally incident light and a setup similar to that shown in Fig. 18.

If a diaphragm is put against the lens O, the structure of the speckle in π' will change when the diaphragm is moved about in its plane. The displacement of the diaphragm introduces a phase factor that is not the same

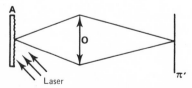

Fig. 17   Speckle in the image π' of a diffuse reflecting object illuminated with a laser.

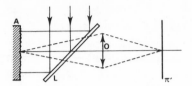

Fig. 18   Normal illumination of a diffuse object A.

for all the points diffracting from the object G. The phase differences in π' between the diffraction patterns emitted from the points of G vary with the position of the diaphragm, and this is responsible for the change in the speckle pattern.

1.8   Changing the Structure of Speckle by Displacing the Plane of Focus*

Consider the transparent diffuse object G illuminated with a laser as shown in Fig. 19. The object might be a ground glass whose image is formed in plane π, by means of the lens O. If the plane of observation π' is moved in a direction perpendicular to the plane, the diffraction patterns of the various points of the ground glass G are modified as seen in Section 1.1. The speckle in π'' itself is modified, because it results from the interference between those diffraction patterns. However, if the displacement δl of the plane is not too great, there will remain some correlation between the speckle patterns in π' and in π''. From Section 1.1, each scattering point of the object G produces as its image a three-dimensional cigar-shaped diffraction pattern whose length is equal to $4\lambda/\alpha^2$ and whose smallest diameter is equal to λ/α, the diameter of the central spot of the diffraction pattern of the lens O. These cigars are oriented as illustrated in Fig. 20. Of course the cigar-shaped patterns are distributed at random in space. If the displacement δl is small compared to the half-length of the cigars, the two speckle patterns in π' and in π'' are correlated. The condition given in Eq. (1.6) must be satisfied.

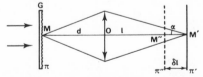

Fig. 19   Changing the structure of speckle by moving the plane of detection.

* See Archbold and Ennos (1972) and Mendez and Roblin (1974).

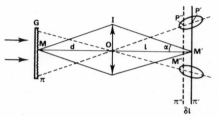

Fig. 20   Small cigar-shaped volumes of diffraction produced in three-dimensional space.

In this case, the speckle in π'' may be deduced from the speckle in π' by the change of scale

$$\frac{M''P''}{M'P'} = \frac{l - \delta l}{l} = 1 - \frac{\delta l}{l} \tag{1.16}$$

Relation (1.16) and Eq. (1.6) show that to increase the scale change, the aperture 2α of the lens O must be reduced.

Instead of displacing the plane π', the object G itself may be moved; the result is the same.

Similar results are also obtained if the object G and plane π' stay where they are, and the lens O is moved parallel to its axis. However, when $d = l$, that is to say, when working under conditions of unit magnification, a displacement δl of the lens has little influence on the path difference $\Delta = (MIM') - (MOM')$. In this case, a change in Δ introduces terms in α with powers higher than second order, which may be neglected, given the approximations already made. As a consequence, for a given scale factor, it will be possible to use an aperture 2α greater than that used for other positions of the elements.

Finally, the speckle in the image of the diffuse object G may be observed for two different magnifications of the image if both the lens O and the plane of observation are moved. The two speckle patterns are identical except for a scale change equal to the ratio of the magnifications (Dzialowski and May, 1976a).

1.9   The Speckle Patterns Produced in the Image of a Diffuse Object under a Change of Wavelength*

Consider again the system shown in Fig. 16. The diffuse object G, say a ground glass, is illuminated with a laser and the speckle pattern in the image

---

* See Mendez and Roblin (1975a) and Tribillon (1974).

plane π' given by the lens O is observed. Without moving anything, the speckle is observed when using a wavelength λ, and then with a wavelength λ'. Chromatic aberration of the lens is assumed to be small enough to have no effect. Under wavelength λ, the speckle pattern is obtained by adding the amplitudes of the diffraction patterns corresponding to the various points of the diffuser G. Of course, the phase variations due to the thickness variations of the ground glass must be taken into account. The same holds true under wavelength λ'.

Consider two arbitrary points M_1 and M_2 of the diffuser, as illustrated in Fig. 21. In the image plane they yield two diffraction patterns centered on the geometrical images M_1' and M_2' corresponding to M_1 and M_2, respectively. Suppose that we are using wavelength λ and that only points M_1 and M_2 are diffracting the light. The resultant amplitude at an arbitrary point M_0' in plane π' is equal to the sum of the amplitudes due to the two diffraction patterns. These are complex amplitudes, because the phase difference due to the difference of thickness at M_1 and M_2 must be taken into account. If n is the index of refraction of the ground glass and δe is equal to the thickness difference at M_1 and M_2, then the path difference at M_0' is equal to $\Delta_G = (n - 1)\delta e$ and the path difference is equal to $\varphi_G = 2\pi\Delta_G/\lambda$. The parameter Δ_G is a characteristic of the diffuse object G, the ground glass in this case. When the observation is made under wavelength λ', the phase difference becomes $\varphi_G' = 2\pi\Delta_G/\lambda'$ if $\delta\lambda = \lambda - \lambda'$ is small enough that Δ_G remains constant. The difference in the phase φ_G due to the wavelength change is equal to

$$2\pi\Delta_G \frac{\delta\lambda}{\lambda^2} \qquad (1.17)$$

The two speckle patterns corresponding to the two wavelengths λ and λ' are practically identical when

$$\Delta_G \frac{\delta\lambda}{\lambda^2} \ll 1 \qquad (1.18)$$

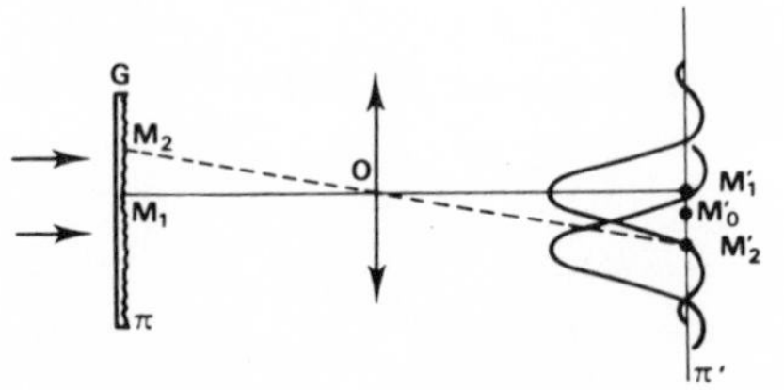

Fig. 21   Changing the structure of speckle in π' by changing the wavelength.

where Δ_G represents an average value for the ground glass under consideration. If this condition is not satisfied, the decorrelation between the two
speckles depends on Δ_G, that is, on the structure of the ground glass.

1.10   White-Light Speckle*

Speckle may be produced artificially from an object illuminated with
any white-light source, if the object is covered by a special type of paint.
Such paints have small transparent beads mixed with small opaque beads.
The paint behaves like a retroreflector: the light beam from the source is
reflected back for the most part toward the source. If the lens forming an
image of the object has sufficient resolution, a large number of bright points
that behave like true speckle may be observed in the image.

When the object is not directly accessible, as in the case of a landscape,
its image may be modulated with an auxiliary diffuser placed against the
photographic plate that is recording the image. The value of this type of
modulation will become apparent when we consider image-processing techniques in Chapter VI.

In certain cases, the objects found in nature have a structure that is sufficiently fine to play the role of speckle grains; for example, sand, grass, piles
of stones, a forest seen from an airplane, and so on. The techniques described
in Chapter VI may be applied directly to such objects, without the need to
modulate the images by means of an auxiliary diffuser.

* See Forno (1975).

CHAPTER II

Speckle Produced at a Finite Distance by a Diffusing Object Illuminated by a Laser

2.1 Fresnel and Fraunhofer Diffraction in Three Dimensions

Consider a screen E_1 with an aperture T illuminated by a parallel beam from a monochromatic point of light (Fig. 22). The aperture T is, for example, a circular aperture. In a plane E_2 located at a finite distance from E_1, we observe the Fresnel diffraction pattern from aperture T. We assume that the diameter of aperture T is small compared to the distance E_1E_2.

We study first the phenomena along the axis of the aperture. The intensity at M is the result of interference between the waves from all the points of

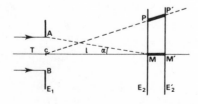

Fig. 22 Variation of the path difference at a finite distance when the plane of observation is moved.

21

aperture T. The path difference at M between the waves from the center C and the edge A of the aperture T is equal to

$$\Delta = MA - MC = \frac{a^2}{2l} \tag{2.1}$$

where a is the radius of the aperture and l is the distance $E_1 E_2$. The path difference at M between the waves from any two points of the aperture T has a value between 0 and Δ, the maximum value of the path difference for the aperture under consideration. The plane of observation is now moved a distance δl. If δl is sufficiently small, the state of interference at M' will be practically the same as that at M. Whatever the luminous intensity at M, its variation along the small segment MM' will be negligible. Equation (2.1) shows that a variation of path difference equal to $a^2 \, \delta l / 2l^2$ corresponds to a displacement δl. For the variation of the state of interference to be negligible along the segment MM', the variation of the path difference must be much smaller than the wavelength λ of the light used.

If $\alpha = a/l$ we have the condition

$$\delta l \ll \frac{2\lambda}{\alpha^2} \tag{2.2}$$

a result already found in Section 1.2.

When the phenomena away from the optical axis are studied, for instance, at a point P a small distance MP away from the axis, practically the same results are obtained. The state of interference remains constant along the segment PP'. This means that the Fresnel diffraction pattern of the intensity in plane E_2' is similar to the Fresnel diffraction pattern of the intensity in plane E_2 and is similarly placed. The intensity distributions in these two planes are similar except for a scale factor $1 - \delta l/l$. If the screen E_2 is moved a distance greater than $2\lambda/\alpha^2$, the Fresnel diffraction pattern is modified. Equation (2.2) shows that the structure of the Fresnel diffraction pattern changes more slowly as the distance $E_1 E_2$ increases.

When the distance from screen E_2 to screen E_1 is great enough, screen E_2 may be moved any distance, but as long as the tolerable displacement δl has a finite value, we are in the domain of Fresnel diffraction. We gradually pass into the domain of Fraunhofer diffraction (diffraction at infinity) by moving screen E_2 away until δl may take practically any value.

If an opaque screen with a great number of very small holes distributed at random and represented by small points on Fig. 23 is now placed in the plane of the aperture T, all these small apertures will send out coherent

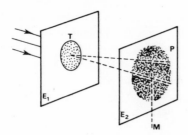

Fig. 23   Fresnel diffraction from a large number of small apertures distributed at random in *T*.

waves and the light diffracted by those small apertures will spread out widely in the plane of observation E_2, because the holes are very small. The interference between those waves produce in plane E_2 a speckle pattern composed of small bright spots.

An arbitrary point *P* of screen E_2 receives a great number of waves whose phases are distributed at random. If screen E_2 is displaced very little, the relative phases of these vibrations will not change very much. The condition for this is that the displacement of screen E_2 satisfy inequality (2.2). Under these conditions, the state of interference remains practically the same as we pass from P_2 to P_2' (Fig. 24). The results may be represented roughly by a set of parallel planes that yield, in a region of space, the distance that the plane of observation may be moved so that the speckle changes only by a scale factor without any appreciable change of structure. As a consequence, the speckle observed in plane E_2 differs from the speckle observed in plane E_2' (or in any plane closer to E_2) only by a scale factor equal to $1 - E_2E_2'/l$,

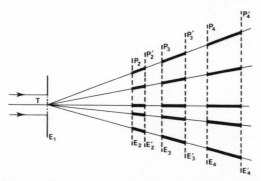

Fig. 24   Speckle observed in two planes so close to each other that the speckle patterns differ only by a scale factor.

where $l = E_1 E_2$. The same result holds true as we pass from plane E_3 to plane E_3' (or to any other plane closer to E_3), from plane E_4 to plane E_4', etc. Of course, the distance that the plane of observation may be moved increases in a continuous fashion with the distance from E_2.

In a region of space where the intervals $E_2 E_2'$, $E_3 E_3'$ have a finite length, the speckle is of the Fresnel type. By increasing the distance of the plane of observation we gradually arrive at a Fraunhofer-type speckle in which the permissible interval between the two planes $E_n E_n'$ increases indefinitely.

2.2   Speckle Produced at a Finite Distance by a Diffuse Object

Depending on whether the object is a diffuse reflecting object or a diffuse transparent object we have Fig. 25a or Fig. 25b. In both cases we study the diffracted light at a distance that is great compared to the size of the aperture containing the diffuse object A. The surface irregularities of object A (Fig. 25a) produce a phenomenon similar to that produced by the small holes in Figs. 23 and 24, and the speckle is observed in the arbitrary plane E_2. The same holds true for the transparent diffuse object of Fig. 25b. In both cases the spread of the diffracted light in an arbitrary plane E_2 is greater if the irregularities on the surface of object A are smaller.

The speckle produced by a diffuse object may be observed easily by means of a ground glass, as shown in Fig. 26. A laser beam is made to diverge by means of a microscope objective O with a magnification of, say, 20, to cover

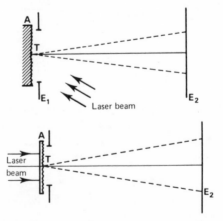

Fig. 25   (a) Speckle produced at a finite distance by a diffuse reflecting object A. (b) Speckle produced at a finite distance by a transparent diffuse object.

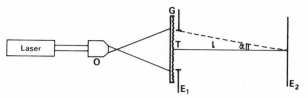

Fig. 26   Apparatus for observing a speckle pattern in E_2.

the ground glass G. Suppose that the ground glass G is limited by a circular aperture T with a diameter $2a$, small relative to the distance l from the plane E_2. The diameter ε of the smallest grains of the speckle in plane E_2 is given by (Gabor, 1970)

$$\varepsilon = \frac{\lambda}{\alpha} \qquad (\alpha = a/l) \tag{2.3}$$

The diameter ε of the grains of the speckle on E_2 does not depend on the structure of the diffuse object G but only on the wavelength λ and on the angle α. By changing the distance l between G and E_2 or the diameter of the aperture T, the scale of the grains of the speckle may be changed at will. For $\lambda = 0.6\ \mu$m and $\alpha = \frac{1}{10}$, the grains have a diameter of the order of 6 μm. The spatial distribution of the speckles depends on the structure of the ground glass. Two ground glasses with the same diameter and located at the same distance from E_2 will yield grains of speckle of the same size, but there will be no correlation between the structures of the two speckle patterns.

If a high-resolution photographic plate is placed at E_2, the speckle may be recorded, and will appear as small black spots on the negative after development. This negative is an excellent diffuser and identical diffusers may be made easily by successfully placing photographic plates at E_2 without changing any of the elements of the setup. Identical diffusers may also be superimposed on the same photographic plate E_2 by taking successive exposures on the same plate and moving the plate in its own plane between each exposure. We shall use this experiment often later on.

2.3   Speckle Produced by a Laterally Displaced Diffusing Object

Consider a diffuser G, for example, a ground glass, illuminated by a distant point source S is derived from a laser (Fig. 27). If the ground glass G and the aperture T are moved together, the speckle produced by G in an

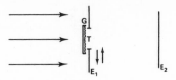

Fig. 27 Lateral displacement of the speckle in E_2 caused by a lateral displacement of the diffuser G.

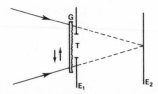

Fig. 28 Lateral displacement of the diffuser G does not change the speckle in E_2.

arbitrary plane E_2 moves with G. Things are different when the geometry of the system is that shown in Fig. 28. The ground glass G is illuminated with a beam that would have converged on the observation screen E_2 if there had been no ground glass in the way. Under these conditions, it is observed that the displacement of ground glass G in its plane (the aperture T is fixed to the ground glass G) has no effect on the speckle observed in plane E_2. The speckle remains stationary. This result is well known in the study of Fraunhofer diffraction: a translation of the aperture in its own plane does not change the intensity distribution in the diffraction pattern produced by the aperture. The same phenomenon is observed with a diffuse reflecting object when the speckle is observed in a plane where the image of the source would form if the diffusing surface were a mirror. In Fig. 28 the displacement of G and T affect only a phase factor in the expression of the complex amplitude on E_2. This phase factor disappears when the complex amplitude is replaced by intensity. Of course it is assumed that the useful part of G limited by T remains covered by the incident beam during the displacement.

2.4   Speckle Produced by a Diffuse Object When the Orientation of the Incident Light Beam Is Changed*

Figure 29 illustrates the principle of this experiment. The ground glass G is illuminated by a beam of parallel rays from a laser and the speckle is

* See Debrus and Grover (1971) and Leger (1976).

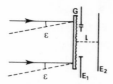

Fig. 29 Lateral displacement of the speckle in E_2 caused by changing the angle of incidence of the light beam on G.

observed in a plane E_2 located at a distance l. The incident beam is initially perpendicular to the ground glass G. We rotate the incident beam by an angle ε and compare the new speckle to that observed when the incident beam was perpendicular to the ground glass. Experiment shows that for a particular ground glass the two speckles remain practically identical for a rotation angle ε that increases as l decreases. The rotation ε of the incident beam produces a simple translation of the speckle on plane E_2. This translation is equal to εl. After a certain value of ε that depends on the roughness of the diffuser, the two speckles become decorrelated.

It is not necessary to illuminate the ground glass with a parallel beam. In the case of Fig. 30, the diffuser is illuminated with a convergent beam from a lens O and a source S_1 derived from a laser. In the absence of the ground glass, the source S_1 would have its image at $S_1{}'$. The speckle is observed on a plane E_2 passing through $S_1{}'$. The source S_1 is now moved to S_2. In the absence of the ground glass, the image $S_1{}'$ would move to $S_2{}'$. The angle of incidence of the chief ray $OS_1{}'$ of the beam varies from 0 to ε. In the presence of the ground glass G, it is observed that the speckle related to $S_2{}'$ is the same as the speckle related to $S_1{}'$. The speckle only moves a distance equal to $S_1{}'S_2{}'$. Of course the displacement of the beam must be sufficiently small, so that the area of the ground glass used in both cases is practically the same.

The decorrelation of speckle as a function of the angle of incidence may also be observed with diffuse reflecting objects, and we shall see that a technique of surface roughness measurement is based on this principle (Section 9.2).

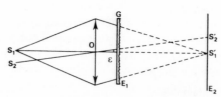

Fig. 30 Translation S_1S_2 of the source produces a translation $S_1{}'S_2{}'$ of the speckle pattern.

2.5 Speckle Produced by a Diffuse Object
under an Axial Translation of the Plane of Observation
or of the Object*

A diffuse object, for instance, a ground glass G, is illuminated with a point source S derived from a laser (Fig. 31). The speckle is observed in a plane E_2 at a distance l from the aperture that limits the ground glass G. From Section 2.1 we obtain in two parallel planes E_2 and E_2' located at a distance δl from each other two speckle patterns differing only by a scale factor, if δl satisfies condition (2.2). For a given wavelength, this condition is a function only of angle α.

Instead of moving the plane of observation, the object G itself may be moved along the axis, and we may look for the resulting changes in the speckle produced in E_2 (Fig. 32). The ground glass is assumed to be infinitely thin and in the plane of aperture T. Consider an incident ray SI that strikes the edge I of aperture T and that is diffracted along IM. The path difference Δ between ray SIM and ray SCM is the maximum path difference corresponding to aperture T in the absence of the diffuser. Let a be the radius of aperture T and d be the distance from source S to aperture T. A simple calculation shows that

$$\Delta = \frac{d + l}{ld} a^2 + \Delta_G \tag{2.4}$$

where Δ_G is the path difference introduced by the ground glass between points I and C. We now move T, that is, the object, a small distance δl, without changing the position of the source S and of the screen E_2. The change in the path difference Δ is

$$\delta\Delta = a^2 \left(\frac{l^2 - d^2}{l^2 d^2}\right) \delta l \tag{2.5}$$

This variation is negligible if $d = l$. Indeed, from Fig. 33 an increase or a decrease of the optical path SI is seen to be compensated by a decrease or an increase of the optical path IM. This means that by giving the diffuse object two distant positions δl apart [condition (2.2)], we shall obtain in plane E_2 two correlated speckle patterns related to each other by a simple scale factor. As in Section 1.8, for a given scale factor, a greater aperture T may be used with this configuration than that that may be used with the optical elements in other positions.

---

* See Dzialowski and May (1976a).

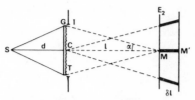

Fig. 31   Changing the structure of speckle by an axial displacement δl of the plane of focus.

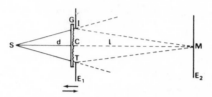

Fig. 32   Changing the structure of the speckle in E_2 by moving the transparent diffuse object G.

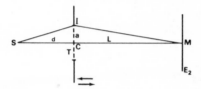

Fig. 33   Changing the path difference at M by moving the aperture T.

Instead of moving the object relative to S and E_2, let us modify the relative positions of S, T, and E_2 in such a way that Δ remains constant. Since diameter $2a$ of the aperture T, which limits the ground glass, is a constant, Eq. (2.4) shows that the two distances d and l must satisfy the relation

$$\frac{d + l}{ld} = \text{const} \tag{2.6}$$

which may be written

$$\frac{1}{d} + \frac{1}{l} = \text{const} \tag{2.7}$$

For any two of those positions, the speckle patterns will be similar with a scale factor equal to l_1/l_2, l_1 and l_2 being two positions of E_2 that satisfy the preceding relation. For diffuse reflecting objects, the configuration shown in Fig. 34 may be used. The source S and the plane of observation E_2 are symmetrical with respect to a beam splitter L inclined at $45°$.

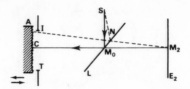

Fig. 34   Changing the structure of the speckle in E_2 by displacing the diffuse reflecting object A.

2.6   Speckle Produced by a Diffuse Object under a Change of Wavelength*

Let the diffuser G, for instance, a transparent diffuse object, be successively illuminated by two wavelengths λ and λ' and let the two speckle patterns produced be observed on plane E_2 (Fig. 35). The random diffuser G has a great number of spatial frequencies, and for a given frequency $1/\zeta_0$ and a wavelength λ, there is a maximum of the light intensity in a direction v given by λ/ζ_0. If the distance E_1E_2 is much greater than the diameter T of the ground glass, a maximum of intensity may be observed at P. If the wavelength is changed, the angle v also changes and the position of maximum intensity moves to P' such that

$$\frac{MP'}{MP} = \frac{v - \delta v}{v} = 1 - \frac{\delta\lambda}{\lambda} \qquad (\delta\lambda = \lambda - \lambda') \tag{2.8}$$

The same reasoning may be applied to each spatial frequency of G. When the wavelength λ is changed to $\lambda' < \lambda$, the speckle becomes smaller, but this contraction does not consist of a simple change of scale in the ratio $1 - \delta\lambda/\lambda$. Two conditions must prevail, which depend on the one hand on the diameter of the diffuser relative to the distance of observation (Section 2.1) and on the other hand on the structure of the ground glass itself (Section 1.9).

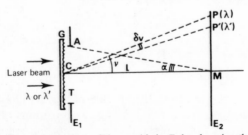

Fig. 35   Changing the structure of the speckle in E_2 by changing the wavelength.

* See Mendez and Roblin (1975).

From Section 2.1, the path difference $\Delta = MA - MC$ is equal to $a^2/2l$. For a wavelength λ, the path difference is equal to $\pi a^2/\lambda l$. Its variation under a change of the wavelength $\delta\lambda$ is equal to $\pi a^2 \delta\lambda/\lambda^2 l$. The influence of a small change of wavelength is negligible if

$$\frac{a^2 \, \delta\lambda}{2l\lambda^2} \ll 1 \tag{2.9}$$

The phase variations caused by the irregularities of the thickness of the ground glass must also be taken into account, as we have done in Section 1.9. This leads to the inequality

$$\Delta_G \frac{\delta\lambda}{\lambda^2} \ll 1 \tag{2.10}$$

where Δ_G represents the average path difference due to the nonuniformity of thickness of the ground glass.

For appropriately chosen wavelengths λ and λ', conditions (2.9) and (2.10) may be satisfied simultaneously. The two speckle patterns obtained on E_2 and corresponding to the respective wavelengths λ and λ' then differ only by a scale factor $1 - \delta\lambda/\lambda$.

2.7 Speckle Produced by a Diffuse Object When the Wavelength and the Position of the Plane of Observation Are Changed*

The speckle is observed with a wavelength λ in plane E_2 and with a wavelength λ' in plane E_2' (Fig. 36). The distance $E_2 E_2' = \delta l$ is chosen so that

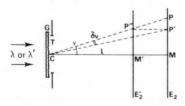

Fig. 36 Changing the structure of the speckle by changing both the plane of observation and the wavelength.

* See Mendez and Roblin (1975a).

TABLE I
THE CORRELATION BETWEEN SPECKLE PATTERNS FOR VARIOUS CASES

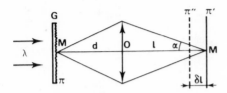

Similar speckle patterns on π' and π'' with the scale ratio $1 - \delta l/l$ if

$$\delta l \ll \frac{2\lambda}{\alpha^2}$$

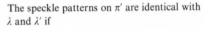

The speckle patterns on π' are identical with λ and λ' if

$$\Delta_G \frac{\delta\lambda}{\lambda^2} \ll 1$$

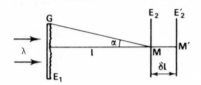

Similar speckle patterns on E_2 and E_2' with the scale ratio $1 - (\delta l/l)$ if

$$\delta l \ll \frac{2\lambda}{\alpha^2}$$

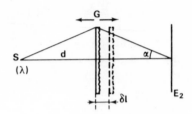

Similar speckle patterns on E_2 with the scale factor $1 - (\delta l/l)$ even with α relatively great, if

$$d = l$$

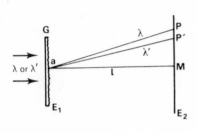

Similar speckle patterns on E_2 with λ and λ' with the scale ratio $1 - (\delta\lambda/\lambda)$ if

$$\frac{a^2\,\delta\lambda}{2l\lambda^2} \ll 1 \cdot \Delta_G \cdot \frac{\delta\lambda}{\lambda^2} \ll 1$$

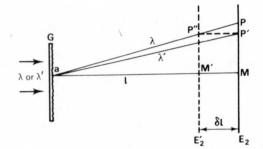

Identical speckle patterns on E_2 with λ and on E_2' with λ' if

$$\delta l = 2\frac{\delta\lambda}{\delta}\frac{a^2\,\delta\lambda}{2l\lambda^2} \ll 1 \cdot \Delta_G \cdot \frac{\delta\lambda}{\lambda^2} \ll 1$$

P'' occupies on E_2' the same position as P' on plane E_2. From Eq. (2.8) this will happen when

$$\delta l = l\frac{\delta\lambda}{\lambda} \tag{2.11}$$

Furthermore, if inequalities (2.9) and (2.10) are satisfied, the speckle observed on E_2' using wavelength λ' is practically identical to the speckle observed on E_2 using wavelength λ.

Table I is a summary of the main results from Chapters I and II regarding the correlation of the observed speckle patterns in different planes with the same wavelength or with two different wavelengths.

2.8 Speckle Produced by a Diffuse Object Itself Illuminated with Another Speckle Pattern*

Consider the experiment illustrated in Fig. 37. A diffuse object A is illuminated, not directly by a laser, but through a ground glass G illuminated by a laser. A speckle pattern from the ground glass G is projected on A with an average angle of incidence θ. Under this condition of illumination, the object A produces a speckle pattern on E_2 that we shall study as the object A is displaced axially, for instance, when it occupies two positions A_1 and A_2 (Fig. 38) separated by a distance δl.

If the object A were illuminated directly with the laser, we have seen that if condition (2.2) is satisfied, a displacement δl will leave the speckle practically unchanged. The grains of the speckle projected on A by G have a diameter of the order of $\lambda/\beta \cos\theta$, where β is the angle under which the

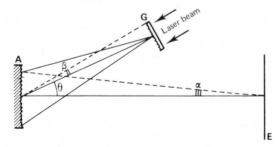

Fig. 37 Illumination of a diffuse reflecting object A by means of an auxiliary speckle produced by G.

* See Eliasson and Mottier (1971) and May and Françon (1976).

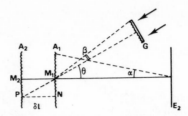

Fig. 38   Axial displacement δl of a diffuse reflecting object illuminated by G.

radius of the ground glass G is seen from the object A. The axial displacement δl produces a lateral displacement M_1N of the speckle pattern produced on the diffuse object by the ground glass G. If M_1N is smaller than the diameter of a grain of the speckle projected on A_1 by the ground glass G, then the speckle in A_1 will be the same as that on A_2. This will be the case when

$$M_1N = \delta l \tan \theta < \frac{\lambda}{\beta \cos \theta} \tag{2.12}$$

that is,

$$\delta l < \frac{\lambda}{\beta \sin \theta} \tag{2.13}$$

If the two conditions (2.2) and (2.13) are satisfied, the speckle patterns produced on E_2 will not be decorrelated. An interesting case, as we shall see later in Chapter VII, is the case where, on the contrary, the axial displacement produces a rapid decorrelation. Condition (2.2) yields values of δl that are relatively large, but not Eq. (2.13) if the angle θ is great enough. A very small displacement δl may then produce decorrelation because condition (2.13) is much more severe than condition (2.2). We obviously have the same phenomena if, instead of observing the speckle at a finite distance, we observe the speckle in the image of a diffuse object with the help of a lens.

CHAPTER III

Interference with Scattered Light

Many experiments done with speckle recorded on photographic plates are experiments of interferometry that may be done with any diffusers. We shall therefore explore first the general principles of interferometry with scattered light, and also the fundamental experiments.

3.1 Historical Background

Although interference with scattered light was known as early as the 17th century, practically no research was done on the subject for almost a hundred years. It was only with the advent of the laser that physicists became interested in these phenomena and in their applications.

Newton (1931) was the first to observe interference with scattered light in the course of experiments done with a concave mirror. He observed rings whose center coincided with the center of curvature of the mirror. He came to the conclusion that the rings were caused by interference phenomena, and he described his observations in the fourth part of his second book, "Optics," under the title "Colours of thick transparent polished plates." Newton introduced the concept of "length of fit," which corresponds to the wavelength of light, and he gave a relationship between the diameters of successive rings, the order of the colors, and the change of diameter of a given ring with the radius of curvature of the reflecting surfaces, or with the thickness of the glass.

Newton's experiments were repeated by le duc de Chaulnes (1755), who accidentally discovered that the intensity of the rings could be increased considerably by breathing on the glass. In his own terms:

> This experiment allowed me to see that the tarnishing of the surface increased the intensity of the phenomenon, and I endeavoured to make it consistent, reliable, and repeatable, because the effect produced by the breath ceased as soon as the moisture evaporated. To this effect, I mixed a drop of milk with ten or twelve drops of water; I spread the mixture on my mirror and let it dry. The white part of the milk was spread out enough to tarnish the mirror to the extent that I needed: I then could observe the phenomenon at leisure. [p. 136]

T. Young (1802) used his theory of light to give an explanation of the phenomenon. He considered the interference of two beams of light: the first, scattered as it entered the glass, then submitted to a specular reflection on the reflecting back surface, exits the plate after an ordinary refraction on the air–glass interface; the second is refracted first on the air–glass surface, then is submitted to a specular reflection on the back surface, and is scattered as it leaves the plate. John Herschel (1830) also studied these phenomena, and Stokes (1851) was the first to give a general mathematical theory. But Burch (1953) was the first to show how interference with scattered light could be applied to interferometry.

3.2   Principles of Interference with Scattered Light

The general principle of these experiments is illustrated in Fig. 39. A plate with parallel faces L is illuminated with a parallel collimated beam of light, that is, by a point source at infinity. In Fig. 39, only one incident ray, SI is shown, and for simplicity, we have assumed normal incidence. The surface AB is made diffuse, and the lower surface is a perfect reflector.

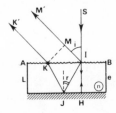

Fig. 39   Interference at infinity of two rays IM' and KK' scattered at I.

Consider the following two paths:

(i)   The incident ray *SI*, refracted normally at *I*, follows the path *IHI* and is scattered at *I* in all directions. Consider, for example, direction *IM'* at an angle *I* to the normal of the plate.

(ii)   The incident ray *SI* is scattered at *I* in all directions. We consider the direction *IJ*, which is such that after a specular reflection at *J*, the ray exits following *KK'* parallel to *MM'*.

These two paths are shown separately in Fig. 40a,b. The scattering by irregularities results in a random change of the phases of the incident rays. Two rays scattered by surface *AB* of plate *L* may interfere even if they are scattered by different points of plate *AB*, but the phase difference changes randomly between two arbitrary rays, and there is no observable interference for the light beam as a whole. Things are different if we consider the rays scattered at the same point of surface *AB*: Two rays such as *SIHIM'* and *SIJKK'* scattered by the same point *I* of the surface *AB* are capable of interference. The same reasoning applies to any point of the surface *AB*, and, as a consequence, to the whole incident beam. Let us calculate the path difference Δ for the two rays just mentioned: *SIHIM'* and *SIJKK'* (Fig. 39 or 40). Let *e* be the thickness of the glass and *n* its index of refraction. The path difference is

$$\Delta = 2n\overline{IH} + \overline{IM} - 2n\overline{IJ} = 2ne(1 - \cos r) \tag{3.1}$$

and if the angles *I* and *r* are small, then

$$\Delta \simeq \frac{ei^2}{n} \tag{3.2}$$

The two rays *SIHIM'* and *SIJKK'* follow two parallel directions *KK'* and *MM'* as they leave the plate *L*. They interfere at infinity, that is, at the focal plane of a lens. Fringes may then be observed by means of the apparatus

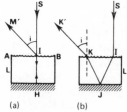

(a)   (b)

Fig. 40   Separation of the paths of two scattered rays.

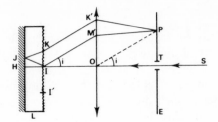

Fig. 41   Observation of circular fringes at *E*.

illustrated in Fig. 41. A screen *E* with a small aperture *T* is put in the focal plane of the lens *O*. The incident beam is a thin (narrow) bundle of rays that passes through the aperture *T*; only one of the incident rays, the ray *SI*, is shown in Fig. 41. The principal axis of the lens *O* is perpendicular to the plate *L*. The arbitrary incident ray *SI* is scattered at *I* either before or after being reflected on the interior face of plate *L*. Just as in Figs. 39 and 40, the two rays that interfere are rays *IHIM'* and *IJKK'* scattered from the same point *I* of surface *AB*. These two rays leave the plate *L* in two parallel directions *KK'* and *IM'*, and they interfere at *P* in the focal plane of lens *O*. Their path difference Δ is given by Eqs. (3.1) and (3.2). The intensity at *P* resulting from the interference of those two rays is given by the classical equation of Fresnel. Assuming equal amplitudes, the intensity is

$$I = I_0 \cos^2\left(\frac{\pi\Delta}{\lambda}\right) \tag{3.3}$$

where I_0 is a constant and λ is the wavelength of the light used. For another point *I'* of the surface *AB*, the phase is different because the scattering introduces a random phase change between the points of surface *AB*. This makes no difference, because for the new point *I'* of *AB*, the same reasoning as before applies. The point *I'* yields two rays that also interfere at *P*, if they are parallel to the preceding rays. They yield the same intensity *I* given by Eq. (3.3), which is added to the intensity due to the interference of the two rays *KK'* and *IM'*. The interference produced by the rays scattered by all the points of *AB* will give the same intensity at *P*. From Eqs. (3.2) and (3.3), the intensity variations on plane *E* depend only on the variable *I*. The interference pattern therefore has a symmetry of revolution about axis *TOI*. The interference fringes are rings centered on *T*. There is a bright ring in direction *i*, that is, at *P* in the focal plane *E* of the lens *O* if

$$\frac{\Delta}{\lambda} = \frac{ei^2}{n\lambda} p \tag{3.4}$$

where p is a whole number. The angular radius of the first bright ring corresponding to $p = 1$ is given by

$$i = \sqrt{\frac{n\lambda}{e}} \tag{3.5}$$

For a glass plate of thickness $e = 0.5$ mm with an index of refraction $n = 1.5$, the first bright ring is seen under an apparent diameter $2i$, which is equal to eight times that of the sun. Certain differences between these rings and the rings at infinity observed with the Michelson interferomenter should be noted. For diffusion rings, the order of interference at the center $(i = 0)$ is always zero whatever the thickness e of the plate. We therefore have a bright center superimposed on aperture T for any value of e. Figure 42 represents the variation of the intensity I as a function of i. The thickness e affects only the diameter of the rings. With the Michelson interferometer, the intensity at the center depends on e. With white light, colors are seen only if the path difference Δ is very small. On the other hand, the diffusion rings are always visible with white light whatever the value of e. They are rings whose center is always white.

Raman *et al.* (1921) have studied diffusion rings by observing the plate L under conditions where the light exits the plate almost parallel with the surface of the plate. In this manner the rays considered are reflected many times, and the rings obtained were similar to multiple beam rings. Fabry and Perot (1897) have observed rings of the same type in transmission with plates having semi-reflecting surfaces with one scattering face. Figure 43 illustrates the principle of this experiment. The plate L in this case is an air space between two semi-reflecting surfaces; this is illuminated by a point source at infinity with an angle of incidence i relative to the normal of plate L. The scattering surface is AB. Consider two parallel rays S_1I_1 and S_2I_2 from a point source at infinity. The ray S_1I_1 is scattered at I_1 and we consider, for example, the scattered ray I_1M_1 which makes an angle α with the normal to the plate. The ray S_2I_2 has a specular reflection at I_2 and at

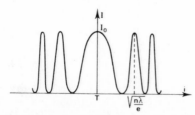

Fig. 42   Intensity distribution in circular fringes.

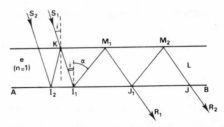

Fig. 43   Interference at infinity of multiple reflected scattered rays.

K, and is then scattered at the same point I_1 of surface AB. After I_1, the ray is reflected many times between the two partially reflecting surfaces. We have represented only two rays J_1R_1 and J_2R_2 and we consider the path difference between the two rays $S_1I_1M_1J_1M_2R_2$ and $S_2I_2KI_1M_1J_1R_1$. The path difference is

$$\Delta = 2e(\cos i - \cos \alpha) \tag{3.6}$$

Taking into account the two extra reflections of the two preceding rays, the path difference Δ remains constant, and so on. The observation could be made by means of the apparatus illustrated in Fig. 44. This setup is similar to that used to observe the rings at infinity from a Fabry–Perot interferometer. Three of the rays arriving at P after multiple reflections are illustrated in Fig. 44. Whereas the Fabry–Perot rings are not visible in white light, things are different here. Whatever the thickness e of the plate L, the path difference of the rays under consideration is always equal to zero if $i = \alpha$. A white ring through the image of the source is observed in the focal plane E of the lens O (the source of white light). This ring shrinks when the angle i become smaller, and if the source is in a direction perpendicular to the plate L, the white ring becomes a white center superimposed on the image of the source.

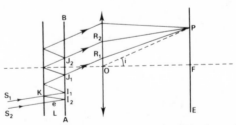

Fig. 44   Observation of circular interference fringes in the case of a multiple wave phenomenon.

3.3   Interference with Two Identical Diffusers

The interference phenomena described in Figs. 39–41 may be considered to be produced by two identical diffusers, the diffuser AB (Fig. 39) and the image of this diffuser reflected on the lower reflecting surface of the plate L. These two diffusers are represented in A_1B_1 and A_2B_2 in Fig. 45. The incident ray scattered at I_1 follows the path I_1KK'. The ray transmitted at I_1 and scattered at I_2 follows the path I_1I_2M'. The path difference between those two rays is given by Eq. (3.2). The same rings at infinity seen in the preceding experiments corresponding to Figs. 39–41 are observed. The angular radius of those rings is given by Eq. (3.5).

This experiment may be carried out with two real identical diffusers obtained by taking photographs of the same speckle pattern on two different photographic plates (Section 2.2). The two diffusers A_1B_1 and A_2B_2 represented in Fig. 46 are two identical diffusers obtained in this manner. In direction i the path difference is

$$\Delta = I_1I_2 - I_1H = ei^2 \tag{3.7}$$

Circular fringes described by Eq. (3.5) when $n = 1$ are observed at infinity. As in the preceding experiments, these rings have a bright center.

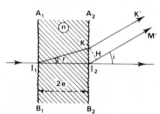

Fig. 45   Analogy of the phenomena described by Figs. 39 and 40, with interference produced by two identical diffusers B_1 and B_2.

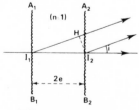

Fig. 46   Path difference $2e - \overline{I_1H}$ between two rays scattered from identical diffusers.

3.4   The Burch Interferometer*

The interferometer described in the following was proposed by Burch to control the quality of a spherical mirror.

The Burch interferometer uses two identical diffusers H_1 and H_2 located near the center of curvature of a spherical mirror M represented in Fig. 47. The two diffusers may be made by the means already described (Section 2.2). An image of H_1 is formed at H_2, after a reflection at M. A narrow beam of parallel rays represented by ray SI passes through H_1, is reflected at C on the mirror, and crosses the diffuser H_2 in the region I' identical to I. At I, part of the beam is scattered in the cone IJJ' and after being reflected on the mirror M, converges to I'. This cone of rays, identified as (1) in the figure, passes through H_2 without being scattered. The beam (2) that passes through H_1 without scattering is reflected at C and is then scattered at H_2 under the same conditions. Rays (1) and (2) leave the diffuser H_2 in the form of two superimposed cones $I'KK'$ and interference between those two waves may be observed. These two waves are obtained in the same manner as previously:

Beam (1) is scattered by H_1 and transmitted by H_2;
Beam (2) is transmitted by H_1 and scattered by H_2.

Beam (2) is reflected at C on a very small area of the mirror and is therefore not affected by the aberrations of the mirror. After being scattered at I', this beam yields a spherical wave Σ_2 centered on I', which serves as a reference wave. Beam (1) yields a wave Σ_1 which is spherical and coincides with Σ_2 only if the mirror M is perfect. If the eye is put behind the surface of the mirror, M appears uniform. If the mirror M has aberrations, the wave Σ_1 is deformed. The interference between waves Σ_1 and Σ_2 reveals the defects

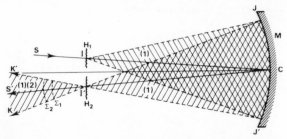

Fig. 47   Burch interferometer with two identical diffusers H_1 and H_2.

* See Burch (1953).

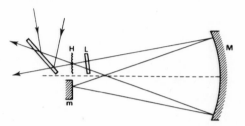

Fig. 48   Burch interferometer with one diffuser H_1.

of mirror M. In Fig. 47 the two waves Σ_1 and Σ_2 have been separated only for the sake of clarity. The fringes represent the lines of equal path differences between the mirror and a perfect spherical mirror corresponding to reference wave Σ_2. The observations may be made with white light, because the path difference is almost zero. When the surface of the mirror is observed, a small luminous spot is seen at C: it corresponds to the beam of light transmitted directly by H_1 and H_2. When the fringes must be recorded on a photographic plate, a small opaque disk is used to mask out this image. This system is very sensitive to vibrations, and to avoid this problem, the second diffuser H_2 may be replaced by a mirror m and the experiment is done in the auto-collimation mode (Fig. 48). The wave Σ_1 comes back and is reflected a second time on the mirror M so that the sensitivity to measure the defects of the mirror M is doubled (Dyson, 1958).

With this system, the reference wave cannot be inclined relative to the deformed wave. This means that we must work with a uniform field. For the fringes to appear, a small glass plate with parallel faces L is interposed at an angle on half of the beam. The plate L is crossed twice, the axial displacement is cancelled, and only the lateral shift between the two waves remains. This shift produces a system of rectilinear parallel fringes if wave Σ_1 is spherical. The deformations of the mirror produce deformations of the wave Σ_1 and, as a consequence, fringes are observed (Scott, 1969; Shoemaker and Murty, 1966).

3.5   Interference Patterns Obtained from Two Laterally Shifted Diffusers*

Consider two ground glasses G_1 and G_2 of the same size, illuminated by a laser. Let us observe the diffraction pattern on a screen E located at a distance from the ground glass that is great relative to the dimensions of

* See May and Françon (1976).

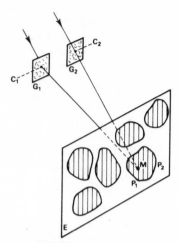

Fig. 49   Interference fringes in speckle grains produced by two arbitrary ground glasses G_1 and G_2.

the diffusers (Fig. 49). The diffraction patterns on screen E may be calculated from the Kirchoff–Fresnel equations, but a simplified explanation may be derived from the following considerations.

Let the ground glass G_1 alone illuminate screen E. Let us consider one grain P_1 of the speckle pattern on this screen. The grains represented in Fig. 49 are of course considerably exaggerated. Grain P_1 is a small luminous spot, inside which the phase is constant. Now if the diffuser G_2 alone illuminated screen E, the speckle pattern produced would have no correlation with the preceding speckle. Suppose that G_2 produced a grain P_2 that occupied the same place as grain P_1 produced by G_1. Inside grain P_2, the phase is constant. When the two diffusers G_1 and G_2 both illuminate screen G at the same time, the two speckle patterns produced on E are coherent, because G_1 and G_2 are illuminated with the same laser beam. The two grains P_1 and P_2 are coherent, and their phases are constant, but not their phase difference, which changes inside the two superimposed grains due to the difference $C_1M - C_2M$, where M is any point inside the grain. Parallel fringes of the Young type are therefore observed inside the grain, the fringe spacing being equal to $\lambda d/C_1C_2$, where d is the distance from screen E to the ground glasses. Since the speckle grains have practically the same amplitudes, the fringes will have maximum contrast.

We have considered the particular case where two speckle grains produced by G_1 and G_2 are superimposed. But the preceding result is general. Any grain of the speckle produced by G_1 and G_2 on E will always be modulated by parallel Young's fringes. All the fringes observed in the grains have the same spacing, but there is no relation between the positions of the fringes when we pass from one grain to another, because the phase distribution of the grains is random. For example, in Fig. 49 there is no reason for the fringes of the grain (P_1, P_2) to be aligned with the fringes of another grain.

If an axial translation is given to either G_1 or G_2, say to G_1, the speckle produced by G_1 on E is modified. The speckle resulting from the interference between the two speckle patterns produced by G_1 and G_2, that is, the speckle pattern observed on E, is also modified. This new speckle pattern has no correlation with the preceding one, but of course each grain is modulated by fringes whose spacing depends on the variation of $C_1M - C_2M$ as before.

If G_1 or G_2 is given a lateral translation, it is not hard to see that the speckle changes along with the fringe spacing inside the grains.

The experimental setup is now modified as shown in Fig. 50. A lens O is placed behind the two diffusers, which are illuminated with a collimated beam, and the diffraction patterns are observed in the focal plane of the lens O. As before, Young's fringes are observed inside the speckle grains. The fringe spacing inside any grain depends on distance C_1C_2 and on the focal length of lens O. This is equal to $\lambda f/C_1C_2$, where f is the focal length of the lens O. There is no relationship between the positions of the fringes when we go from one grain to another. This phenomenon may be compared to the diffraction pattern produced by two identical apertures. Let the two diffusers G_1 and G_2 of Fig. 50 be replaced by two aperture with the same

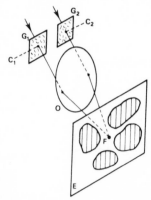

Fig. 50 Interference fringes in speckle grains produced by two arbitrary ground glasses in the focal plane of a lens O.

dimensions. In the focal plane F one observes the diffraction pattern of one of the apertures modulated by the interference fringes with a spacing equal to $\lambda f/C_1C_2$, where C_1 and C_2 are two corresponding points of the two apertures. When the two apertures are replaced by the two diffusers, having obviously different structures, there is a random repetition in focal plane F of the phenomena observed before with the two apertures.

Let either G_1 or G_2, say G_1, be translated axially. The structure of the speckle pattern produced by G_1 in the focal plane F of lens O does not change. The same is true for the speckle produced by G_1 and G_2. This means that the relative positions of the speckle grains do not change. This axial translation only introduces a quadratic phase factor. Inside each grain these rectilinear fringes become segments of circular fringes. These fringes are centered on point P_0 determined by passing a straight line parallel to C_1C_2 through O (Fig. 51). Of course there is no relationship between the positions of these fringes as we pass from one grain to another.

Let G_1 or G_2, say G_1, be translated laterally. The speckle pattern produced by G_1 does not change, that is, the relative positions of the grains in the speckle pattern observed in the focal plane do not change. The lateral translation only introduces a linear phase factor. Inside each grain the fringe spacing is modified. If we are in the case of Fig. 50, the change in the fringe spacing is the same for all the grains. In the case corresponding to Fig. 51 (the two diffusers are not in the same plane), the center P_0 is translated in a way that is easy to determine with the help of Fig. 51.

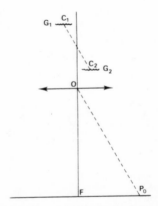

Fig. 51 Two ground glasses G_1 amd G_2 yield in the focal plane of lens O portions of circular fringes in the speckle grains.

3.6   Interference Patterns Obtained from Two Different Axially Shifted Diffusers*

This case is related to the one that we have just studied (Fig. 51). Consider two diffusers G_1 and G_2 along an incident collimated beam from a laser (Fig. 52). First of all, it is assumed that the diffusers G_1 and G_2 are not ground glasses, but partial scatterers of the type described in Section 3.3 and used to carry out the experiment of Fig. 46. These are photographs of speckle patterns. In the experiment corresponding to Fig. 46, two photographs of the same speckle pattern were used, that is, two identical diffusers. This time, two different diffusers G_1 and G_2 are used. They are illuminated with a collimated beam, and the diffraction pattern is observed in the focal plane of lens O (Fig. 52). At the focal point F there is an image of the source that

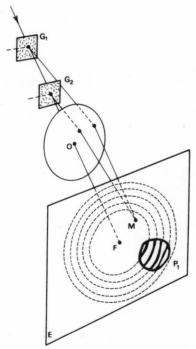

Fig. 52   Interference fringes produced in the focal plane of lens O by two axially shifted diffusers G_1 and G_2.

* See Dainty (1976).

corresponds to the unscattered rays transmitted directly by G_1 and G_2. It is not this image that is of interest, but only the field surrounding it. We may reason along the lines of Section 3.5. Interference fringes are observed only in the speckle grains. Here the fringes are circular fringes centered on the focal point F, but there is no relationship among the positions of the fringes between the various grains. In Fig. 52 the circular fringes that will be obtained when the two diffusers G_1 and G_2 are identical have been represented by the broken lines. In an arbitrary plane P_1 the spatial distribution of the fringes is the same as that of the fictitious broken fringes that pass through this region, but there is a random shift between those two systems of fringes.

If either of the two diffusers G_1 or G_2 is given an axial translation, it is seen from the preceding that the speckle pattern does not change, that is to say that the spatial distribution of the grains remains the same, but the fringes inside the grains change. This is due to the variation in the quadratic phase factor. The radii of these fringes may increase or decrease depending on whether G_1 is moved away or towards G_2, and this happens for all the grains.

The results are the same if the two preceding diffusers are replaced by two ground glasses. The only difference is that the phenomena may only be observed in the regions of E where the rays scattered by G_1 pass outside of G_2. This is the case of point M of Fig. 52.

All these results will be useful in understanding the phenomena described in Chapter VII.

CHAPTER IV

Interference Patterns Produced by Photographic Superposition of Laterally Shifted Speckle Patterns

4.1 Amplitude Transmitted by a Photographic Plate after Development

The experiments that follow were produced with photographs of speckle patterns; it is therefore important to understand the elementary properties of photographic plates after development. A possible experiment is that described by Fig. 53. A ground glass G limited by an aperture T is illuminated with a laser and the speckle pattern is recorded on a photographic plate H located at a distance l from the ground glass. The light from the laser is

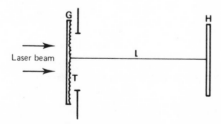

Fig. 53 Photographic recording at H of the speckle pattern produced by G.

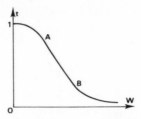

Fig. 54   Amplitude transmittance of a negative after development.

polarized, and except where indicated, it is assumed in what follows that the polarization of the light does not change during the experiments.

The speckle at H is described by a function $D(\eta, \zeta)$, which represents the intensity of the light on H (see Section 1.5). If T is the exposure time, the plate H receives an energy $W = T \times D(\eta, \zeta)$. We are interested in the relation between the energy W received on the plate and the amplitude transmission t after development. The curve representing the function $t = f(W)$ is given in Fig. 54. This curve has a linear region AB: in this region the transmitted amplitude t is proportional to the incident energy during the exposure. If there is only one exposure, or a number of exposures with equal exposure times, it is not necessary to consider explicitly the exposure times. If we are in the linear region AB of the curve, a linear relation between t and the intensity $D(\eta, \zeta)$ may be written. It may be noted that the linear part AB of Fig. 54 corresponds to the region of underexposure of the H-D curve of the emulsion.

Consider the experiment of Fig. 53, and assume a single exposure. The conditions are such that the intensity variations of the speckle, that is, the variations of the function $D(\eta, \zeta)$, are such that the corresponding variations of the amplitude transmitance after development are in the linear part of Fig. 54. Under these conditions, the transmitted amplitude t of the negative H is given by

$$t = a - bD(\eta, \zeta) \tag{4.1}$$

where a and b are two characteristic constants of the photographic emulsion used.

4.2   The Fundamental Experiment of Burch and Tokarski*

Consider again the experiment of Fig. 53, but instead of a single exposure, two equal exposures are made, and the photographic plate H is shifted

* See Goodman (1976).

between the two exposures. The plate is shifted for example a distance ζ_0 in the direction of axis ζ. The recorded intensity is the sum of the recorded intensities in each exposure, and is equal to

$$D(\eta, \zeta) + D(\eta, \zeta - \zeta_0) \tag{4.2}$$

Now a translation is equivalent to convolution with a delta function (see Section 1.5), and so the intensity may be written in the form

$$D(\eta, \zeta) \otimes [\delta(\eta, \zeta) + \delta(\eta, \zeta - \zeta_0)] \tag{4.3}$$

where $\delta(\eta, \zeta)$ is a delta function centered on (η, ζ). After development, the amplitude t transmitted by the negative H is

$$t = a - bD(\eta, \zeta) \otimes [\delta(\eta, \zeta) + \delta(\eta, \zeta - \zeta_0)] \tag{4.4}$$

The spectrum of the negative is then studied with the apparatus illustrated in Fig. 55. The negative H is illuminated with a collimated beam of light from a monochromatic point source with a wavelength λ and located at infinity. In the focal plane F of the lens O, the spectrum of the negative H is observed, that is, the Fourier transform of the amplitude t transmitted by H. This transform $\tilde{t}$ may be written

$$\tilde{t}(u, v) = a\delta(u, v) - b\tilde{D}(u, v)[1 + \exp(j\pi v\zeta_0/\lambda)] \tag{4.5}$$

where the tilde represents the Fourier transform and (u, v) are the angular coordinates of a point in the focal plane of lens O.

The first term $a\delta(u, v)$ on the right-hand side of Eq. (4.5) represents the image of the point source at infinity when diffraction effects are neglected. This image is located at the focus F and is very small. In the second term, except for constant b, the Fourier transform $\tilde{D}(u, v)$ of $D(\eta, \zeta)$ is modulated by $1 + \exp(j\pi v\zeta_0/\lambda)$. The diffuser $D(\eta, \zeta)$ is very fine and its transform $\tilde{D}(u, v)$ spreads out considerably in the focal plane of lens O. Like $D(\eta, \zeta)$, the transform $\tilde{D}(u, v)$ has the structure of speckle. If the image of the light source at

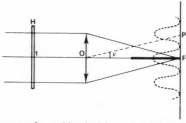

Fig. 55 Fourier spectrum of two identical laterally shifted speckle patterns (Burch and Tokarski, 1968).

F is neglected, the light intensity in the rest of the focal plane is proportional to

$$I = |\tilde{D}(u, v)|^2 \, |1 + \exp(j\pi v\zeta_0/\lambda)| = |\tilde{D}(u, v)|^2 \cos^2\!\left(\frac{\pi v\zeta_0}{\lambda}\right) \qquad (4.6)$$

The diffuse background $|\tilde{D}(u, v)|^2$ is modulated by $\cos^2(\pi v\zeta_0/\lambda)$, which represents Young's fringes. The angular distance between two bright (or dark) consecutive fringes is equal to λ/ζ_0. For example, for a translation ζ_0 equal to 20 μm, the angular distance between two consecutive bright fringes is equal to $1°42'$, that is, more than three times the apparent diameter of the sun.

The foregoing expressions show that the same results are obtained as in Section 1.5. Indeed, the speckle on the negative H is composed of small dark spots, and from Babinet's theorem, the spectrum is identical to that that would have been obtained from the complementary screen consisting of small holes in an opaque screen (except at F). In the preceding experiment, the two exposure times are equal, and the Young's fringes have the maximum contrast, that is, the minima have values equal to zero.

When the two exposure times are unequal, the recording is described by the expression

$$D(\eta, \zeta) \otimes [B_1\delta(\eta, \zeta) + B_2\delta(\eta, \zeta - \zeta_0)] \qquad (4.7)$$

where B_1 and B_2 are proportional to the two exposure times. It is easy to show that expression (4.6) yields, except for a constant factor,

$$I = |\tilde{D}(u, v)|^2\left[B_1^2 + B_2^2 + 2B_1B_2 \cos\!\left(\frac{\pi v\zeta_0}{\lambda}\right)\right] \qquad (4.8)$$

Let I_{max} be the intensity of the bright fringes:

$$I_{max} = |\tilde{D}(u, v)|^2(B_1 + B_2)^2 \qquad (4.9)$$

and I_{min} the intensity of the dark fringes:

$$I_{min} = |\tilde{D}(u, v)|^2(B_1 - B_2)^2 \qquad (4.10)$$

The contrast of the fringes is equal to

$$\gamma = \frac{I_{max} - I_{min}}{I_{max} + I_{min}} = \frac{2B_1B_2}{B_1^2 + B_2^2} \qquad (4.11)$$

The greater the difference between B_1 and B_2, the lower the contrast.

The contrast is obviously greatest when $B_1 = B_2$, that is, when the two exposure times are equal.

In the experiment of Fig. 55, the negative H may be illuminated with a point source of white light. In this case, the classical colors of white-light interference phenomena will be observed.

Note: In the experiment of Fig. 53, the photographic plate H was translated between the two exposures. We may also leave plate H stationary, and translate the ground glass G. The results will be the same except if the ground glass G is illuminated with a convergent beam. Indeed, it was seen in Section 2.3 that if the beam converges on H when the ground glass is absent, there is no displacement of the speckle on H when the ground glass is present.

4.3 The Superposition of a Series of Successive Exposures on the Same Holographic Plate

Let us make a series of successive exposures with equal exposure times and equal translations in the same direction between each exposure. If $N + 1$ exposures are made, the recording will be

$$D(\eta, \zeta) \otimes \sum_{n=0}^{N} \delta(\eta, \zeta - n\zeta_0) \tag{4.12}$$

After development, the negative is illuminated as in the experiment of Fig. 55. At any point (u, v) in the focal plane of lens O, the amplitude is given by the Fourier transform of Eq. (4.12). Except for the image of the source at F and a constant, factor, the intensity is equal to

$$I = |\tilde{D}(u, v)|^2 \left\{ \frac{\sin[(N + 1)\pi v\zeta_0/\lambda]}{\sin(\pi v\zeta_0/\lambda)} \right\}^2 \tag{4.13}$$

The spectrum observed in the focal plane of lens O is the same as that from a grating with a period equal to ζ_0 and a number of bars equal to $N + 1$. Between two principal maxima, there are $N - 1$ secondary maxima (Fig. 56).

If the number $N + 1$ of exposures is great enough, the recording may be represented by the simple expression

$$D(\eta, \zeta) \otimes \text{comb}\left(\frac{\zeta}{\zeta_0}\right) \tag{4.14}$$

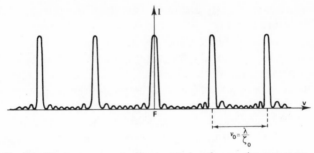

Fig. 56 Spectrum corresponding to a series of successive exposures.

Fig. 57 Representation of a spectrum by a Dirac comb.

The intensity spectrum in the focal plane of lens O shows in Fig. 57 is

$$I = |\tilde{D}(u, v)|^2 \, \text{comb}\left(\frac{v}{v_0}\right) \tag{4.15}$$

where $v_0 = \lambda/\zeta_0$. Of course these results are rather theoretical, because the number of possible exposures is limited by the dynamic range of the photographic emulsion.

It may be shown that the secondary maxima that appears in Fig. 56 can be suppressed if the exposure times B_1, B_2, B_3, etc., are proportional to the binomial coefficients $C_N{}^P$. In order to show this, start from a central exposure and make successive exposures while translating the photographic plate symmetrically relative to the central exposure. The secondary maxima will be suppressed if the modulation factor is an expression such as $\cos^2(\pi v \zeta_0/\lambda)$, $\cos^4(\pi v \zeta_0/\lambda)$, $\cos^6(\pi v \zeta_0/\lambda)$, that is, $\cos^N(\pi v \zeta_0/\lambda)$, where N is even. This yields

$$2^N \cos^N\left(\frac{\pi v \zeta_0}{\lambda}\right) = \sum_{n=0}^{N} C_N{}^P \exp[j(\tfrac{1}{2}N - n)2\pi v \zeta_0/\lambda] \tag{4.16}$$

The exposure times and the required translations are given in Table II.

TABLE II

Number of exposures	Modulation factors	Translations and exposure times				
3	$\cos^4\left(\dfrac{\pi v \zeta_0}{\lambda}\right)$		$-\zeta_0$ $B_2 = 1$	0 $B_1 = 2$	$+\zeta_0$ $B_3 = 1$	
5	$\cos^8\left(\dfrac{\pi v \zeta_0}{\lambda}\right)$	$-2\zeta_0$ $B_4 = 1$	$-\zeta_0$ $B_2 = 4$	0 $B_1 = 6$	$+\zeta_0$ $B_3 = 4$	$+2\zeta_0$ $B_5 = 1$

For the reasons previously given, it is difficult to have more than $N + 1 = 7$ exposures.

4.4   Simultaneous Recordings with Birefringent Plates*

The experiment of Burch and Tokarski may be done with a single exposure by shifting the speckle with a birefringent plate. Consider a uniaxial bire-fringent plate with parallel faces Q cut at $45°$ to the optical axis (Fig. 58). For a quartz plate, the lateral shift ζ_0 is equal to

$$\zeta_0 = 5 \times 10^{-3} e \tag{4.17}$$

where e is the thickness of the quartz plate.

For $\zeta_0 = 20 \ \mu\text{m}$, the thickness e is equal to 4 mm. Because it is the in-tensities of the speckle patterns that are added in the experiment of Burch and Tokarski, the plate Q must be used without a polarizer. The two speckle patterns, one corresponding to the ordinary wave, and the other to the extraordinary wave, are added in intensity. There is a slight additional axial

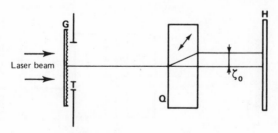

Fig. 58   Use of a birefringent plate.

* See Françon *et al.* (1976).

shift between the speckle corresponding to the ordinary wave and the speckle corresponding to the extraordinary wave. This axial shift is equal to

$$\delta l = 0.004e \qquad (4.18)$$

From Eq. (2.2), these two speckle patterns are practically identical on H if

$$\delta l = 0.004e \leq \frac{2\lambda}{\alpha^2} \qquad (4.19)$$

where 2α is the angle under which the ground glass is seen from H. For a thickness $e = 7$ mm, α must not in theory be greater than about $\frac{1}{5}$.

In certain cases (Chapter VI), it may be of advantage to make three or five exposures in order to have finer fringes. The apparatus illustrated in Fig. 59 allows three images to be recorded with two exposures. Two birefringent plates with parallel faces Q_1 and Q_2 oriented at 45° as shown in Fig. 59 are interposed between G and the photographic plate H. The principal planes of these two plates are parallel, and a half-wave plate is placed between Q_1 and Q_2. The axes of the half-wave plate are oriented at 45° relative to the principal planes of plates Q_1 and Q_2. The ordinary ray in the first plate becomes extraordinary in the second and vice versa. The shift between the ordinary and the extraordinary rays produced by the first plate Q_1 is doubled by the second plate Q_2. The two emerging rays are symmetrical with respect to the ordinary ray in the first plate Q_1, that is to say, $CP_1 = CP_2$. Because these two rays are incoherent, there will be on H two identical speckle patterns shifted relative to each other by $P_1P_2 = 2\zeta_0$, where ζ_0 has a value given by Eq. (4.17) in the case of quartz plates.

If the half-wave plate is rotated through 45°, the shift produced by Q_1 is cancelled by Q_2 (Fig. 60). There is only one speckle pattern on H. This speckle pattern is identical to the two preceding speckle patterns. In order

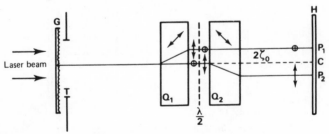

Fig. 59 Simultaneous production of three incoherent images with intensity proportional to 1, 2, 1 (the observation of the two images with unit intensity).

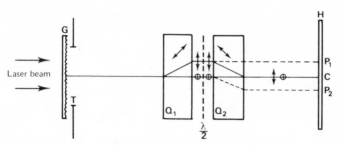

Fig. 60   Obtaining an image with twice the intensity.

to pass from the speckle pattern corresponding to Fig. 60 to the two speckle patterns corresponding to Fig. 59, two equal translations in opposite directions are required. It may be seen that the speckle pattern obtained on H in the case illustrated by Fig. 60 is twice as bright as the two speckle patterns obtained in the case of Fig. 59. These are exactly the conditions corresponding to the table of Section 4.3 (first line, three exposures). As a consequence, if two successive exposures are made on H, one corresponding to Fig. 59, the other to Fig. 60 (without moving H), the intensity in the spectrum of the negative after development will yield fringes of the type $\cos^4(\pi v \zeta_0/\lambda)$.

In many phenomena (Chapter VII), it is necessary to study the changes of a speckle pattern between two times t_1 and t_2. Two speckle patterns must therefore be recorded, one at time t_1 and the other at time t_2. The exposures must be superimposed on plate H. The preceding apparatus is of interest because with only two exposures, finer fringes are obtained in the spectrum of the negative, which means, as we shall see, that the quality of the observed phenomena is improved.

If additional birefringent plates are added, it is possible to obtain five images, seven images, etc., with shifts and intensities that satisfy the conditions given in Table II (Section 4.3). These systems are more complicated and less interesting than the three-image apparatus, which is sufficient in almost all cases.

The preceding birefringent plates with parallel faces may be replaced by prismatic plates, which as a whole form plates with parallel faces as shown in Fig. 61. In this case an angular shift is obtained. When the axes of the half-wave plate are parallel to the axes of the prismatic plates, the image is in direction IC. If the half-wave plate is rotated through $45°$, two images in directions IP_1 and IP_2, symmetrical relative to IC, are obtained. The angular shift α is the same as that of a birefringent Wallaston prism:

$$\alpha = 2(n_o - n_e)\tan\theta \qquad (4.20)$$

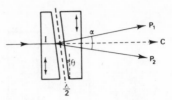

Fig. 61 Technique to obtain three incoherent images with a Wollaston prism.

where n_o and n_e are the ordinary and the extraordinary indices of the bire-fringent material used. For a quartz prism with an angle $\theta = 5°$, we have $\alpha = 6'$.

4.5   Recording with a Continuous Displacement of the Photographic Plate*

Consider the experiment of Fig. 53, where the ground glass G produces a speckle pattern on photographic plate H. During the exposure, plate H is moved in any direction at a constant speed. Let ζ_0 be the distance moved during the displacement. The continuous translation of the plate H during the exposure is equivalent to a convolution of the speckle pattern with a rectangular function having a width ζ_0. The recording may therefore be written

$$D(\eta, \zeta) \otimes \text{Rect}\left(\frac{\zeta}{\zeta_0}\right) \tag{4.21}$$

The amplitude t transmitted by the negative after development is

$$t = a - b\left[D(\eta, \zeta) \otimes \text{Rect}\left(\frac{\zeta}{\zeta_0}\right)\right] \tag{4.22}$$

If the spectrum of the negative is observed as in Fig. 55, Eq. (4.5) becomes

$$\tilde{t}(u, v) = a\delta(u, v) - b\tilde{D}(u, v)\frac{\sin(\pi v\zeta_0/\lambda)}{\pi v\zeta_0/\lambda} \tag{4.23}$$

If the image of the light source at F is neglected, the light intensity in the focal plane of lens O is given by

$$I = |\tilde{D}(u, v)|^2 \left(\frac{\sin(\pi v\zeta_0/\lambda)}{\pi v\zeta_0/\lambda}\right)^2 \tag{4.24}$$

* See Tiziani (1972a).

The diffuse background $|\tilde{D}(u, v)|^2$ is modulated by an intensity function identical to that that is observed in the case of diffraction by a slit whose width ζ_0 is equal to the translation of the plate H during the exposure.

If during the exposure plate H is given a movement represented by the function $f(\zeta_0)$, the Fourier transform $\tilde{f}(v)$ of the function $f(\zeta_0)$ may be observed in the spectrum of the negative.

4.6   Recordings Obtained When the Orientation of the Incident Beam Is Changed

Consider the experiment of Fig. 62. Two successive recordings are made on plate H while the orientation of the incident beam on the ground glass is changed. During the first exposure, the beam is perpendicular to the ground glass, and the beam is given a rotation ε before the second exposure. It was seen in Section 2.4 that if l is sufficiently small, the rotation ε of the

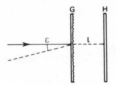

Fig. 62   Photographic recording of two speckle patterns shifted by changing the orientation of the incident beam.

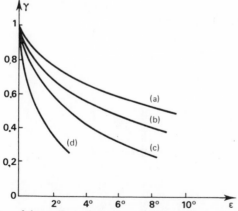

Fig. 63   Variations of the contrast of the fringes in the spectrum of the recording obtained as shown in Fig. 62. (a) $l = 1.5$ mm; (b) $l = 3.0$ mm; (c) $l = 5.8$ mm; (d) $l = 11.5$ mm.

incident beam results simply in a translation of the speckle pattern on the plate *H*. When two exposures are made under these conditions, the spectrum of the negative *H* after development will yield Young's fringes as in the experiment of Burch and Tokarski. If *l* (or ε) is increased, this will introduce a decorrelation between the two recorded speckle patterns, which will result in a lowering of the contrast of the observed fringes in the spectrum. Figure 63 shows the contrast of the fringes when expression (4.1) is used as a definition of the contrast γ. The same experiment may be made with a diffuse reflecting object.

4.7 Recordings Made with Polarized Light*

Let us repeat the experiment described in Fig. 53, with incident light linearly polarized by means of a polarizer $\mathscr{P}$ shown in Fig. 64. Two successive exposures are made, and the polarizer $\mathscr{P}$ is rotated through an angle θ between the two exposures. The spectrum of the negative is then observed as in the preceding experiments with the apparatus shown in Fig. 55. If the diffuse object *G* is a ground glass, the contrast of the fringes is observed to decrease as the angle θ increases. It becomes equal to zero when $\theta = \pi/2$. When the two exposures correspond to two perpendicular directions of polarization, the two speckle patterns on *H* are not correlated. Similar results are obtained with other diffuse objects, which, like ground glass, do not depolarize the light very much. With magnesia, the scattered light is completely depolarized, and the speckle pattern does not change with the orientation of the incident light.

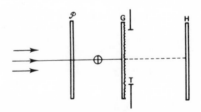

Fig. 64 Correlation between the speckle patterns at *H* when the polarization of the incident light is changed.

* See Chandley and Welford (1975).

4.8   Recordings Obtained When the Orientation
of the Diffuse Object Is Changed*

Consider first a plane diffuse reflecting object that is rotated through a small angle θ about an axis through the object. The incident illumination is perpendicular to the object, and the speckle is observed in a direction close to the normal. The speckle is observed to move as if the object were a mirror. At a distance D from the object, the speckle is shifted a distance equal to $2\theta D$. This is also true at infinity, that is, if the speckle is observed in the focal plane of a lens. This result may be explained by remembering that a diffuse surface may be considered to result from the superposition of a large number of reflection gratings with random orientations and periods. Any rotation θ of the grating causes the spectra to rotate through an angle 2θ.

The phenomena are different for a diffuse transparent object. Indeed, for a transmission grating, it is known that a small rotation of the grating is equivalent to a change in the period of the grating. This results in a small change of the position and the spread of the spectra.

Consider a transmitting diffuser G such as a ground glass shown in Fig. 65. Two successive exposures are made on the same photographic plate H, and between the two exposures the orientation of the diffuser G is changed. The diffuser G is rotated about an axis O perpendicular to the plane of the page, and in the plane of the diffuser. The diffuser is rotated through a small angle θ, and successively occupies the two positions G_1 and G_2. After the two exposures, the negative itself and not its spectrum is observed. The shift between the two speckle patterns recorded on H is seen to depend on the point considered on H. There is a circumference on H along which the shift between the two speckle patterns is equal to zero. This circumference has a

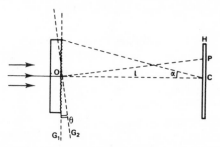

Fig. 65   Recordings made at H when the orientation of a diffuse transparent object is changed.

* See Mendez and Roblin (1975b).

radius equal to $l\theta/2$, and it is centered on point P, located at a distance $l\theta/2$ from point C. The existence of this phenomenon may be observed in the spectrum of negative H by means of the apparatus shown in Fig. 55. An opaque screen with a small aperture is used to mask negative H. When the aperture is on the circumference corresponding to zero shift, there are no fringes in the spectrum. When the aperture is in other regions of H, fringes with various spacings are seen.

The contrast of the fringes may be calculated from the Kirchhoff–Fresnel equation. The contrast depends on the structure of the diffuser and on the angle of rotation θ. If α is assumed to be small, and if Δ is the distance between any point of the diffuser and the average plane of the diffuser surface, the contrast of the fringes is equal to zero when

$$\Delta = \frac{\lambda n}{\theta^2} \tag{4.25}$$

where n is the index of refraction of the diffuser. If the axis of rotation of the diffuser is not in the plane of the diffuser but at a small distance d from this plane, the circumference of zero shift remains centered on point P located at a distance $l\theta/2$ from point C, but its radius R becomes equal to

$$R = \frac{l\theta}{2}\left(1 + \frac{8d}{l\theta^2}\right)^{1/2} \tag{4.26}$$

This phenomenon is also present in the case of a diffuse reflecting object considered at the beginning of this section, but it is hidden by the much more important effect of the rotation of the speckle.

4.9   Recordings Made with More Than One Wavelength

Consider a diffuser G illuminated by light having a wavelength λ_1 as shown in Fig. 66. A dispersive system $\mathscr{D}$, shown here as a prism for simplicity, is interposed, and the resulting speckle pattern is recorded on a photographic plate H. The diffuser G is then illuminated with another wavelength λ_2. A second speckle pattern is obtained on H, and it is shifted by a distance P_1P_2 relative to the speckle pattern of the first recording. The difference between the wavelengths λ_1 and λ_2 is assumed to be small, and the energy output from the light source is the same in both cases. The difference $\lambda_1 - \lambda_2$ is assumed to be small enough that the phase variations caused by the roughness of the diffuser do not change when we pass from λ_1 to λ_2. The two speckle patterns corresponding to the two wavelengths λ_1 and λ_2 are then practically

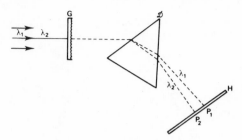

Fig. 66   Recordings made at H when the wavelength is changed.

identical, and the conditions are the same as those of the experiment of Burch and Tokarski, with $\zeta_0 = P_1 P_2$, described in Section 4.2. The translation of the photographic plate H between the two exposures is replaced here by the deviation of the prism, which is different for the two wavelengths. After development, the spectrum of the negative is observed with wavelength λ_1 or λ_2 as in Fig. 55. The spacing of the fringes yields the difference $\lambda_1 - \lambda_2$, if the apparatus is previously calibrated with known wavelengths. In this experiment, the interference fringes observed in the spectrum of negative H may be said to represent the Fourier transform of the autocorrelation function of the spectrum of the source. Because the spectrum $I(\lambda)$ of the source consists in this case of two lines with the same intensity, the autocorrelation function of $I(\lambda)$ consists of three signals with relative amplitudes 1,2,1, which yields a Fourier transform equal to $\cos^2 x$.

This result is general, and the experiment of Fig. 65 allows the Fourier transform of the autocorrelation function of the spectrum $I(\lambda)$ of the light source illuminating the diffuser G to be obtained, but the superposition of speckle patterns corresponding to many wavelengths causes a loss of information, and the signal-to-noise ratio becomes weak.

CHAPTER V

Interference Patterns Produced
by Photographic Superposition
of Axially Shifted
Speckle Patterns

5.1 Circular Interference Fringes Produced by Two Successive Recordings on the Same Photographic Plate*

An arbitrary diffuse object, for example, a ground glass G, is illuminated with a point source S derived from a laser and the speckle pattern is recorded on the holographic plate H as shown in Fig. 67. Two successive exposures are made on plate H, which is translated between exposures in a direction perpendicular to its plane. If the displacement of the photographic plate is

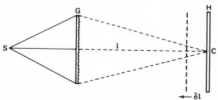

Fig. 67 Recording two speckle patterns that are identical except for a scale factor in two parallel planes.

* See Mendez and Roblin (1974).

65

not too great, that is, if δl satisfies condition (2.2), two speckle patterns differing only by a scale factor $1 - \delta l/l$ are recorded on H (Section 2.5 and Table I).

Consider the properties of the negative after development (Fig. 68). At a distance y from the center C of the photographic plate, there is a distance $y \, \delta l/l$ between two corresponding elements of the speckle patterns. Let the negative be illuminated with a parallel beam of light. If Δ is the path difference of the speckles at a point P located at any distance l on the perpendicular to the plate at C, then

$$\Delta = y \frac{\delta l}{l} \alpha = y^2 \frac{\delta l}{lL} \tag{5.1}$$

We now interpose an opaque screen E_1 with a very small aperture that coincides with T, as shown in Fig. 69, and observe the phenomena on a screen E_2 located behind E_1. The intensity at T resulting from the interference of the waves sent out from the two similar points located at a distance y from C is equal to

$$I = I_0 \cos^2\left(\frac{\pi \Delta}{\lambda}\right) = I_0 \cos^2\left(m \frac{y^2}{\lambda}\right) \tag{5.2a}$$

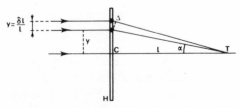

Fig. 68 Path difference at T between two similar elements of the two speckle patterns recorded on H.

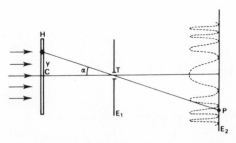

Fig. 69 Observation of the circular interference fringes.

or

$$m = \frac{\pi \, \delta l}{lL} = \text{const} \tag{5.2b}$$

Except for diffraction effects and a constant factor, this is also equal to the light intensity at P on screen E_2. Because of the circular symmetry, circular rings are observed on screen E_2. In white light, these rings behave like Newton's rings with white centers. For a wavelength λ, the radius of the pth ring is equal to

$$y = \sqrt{\frac{\lambda l L}{\delta l}} \sqrt{p} \tag{5.3}$$

Instead of observing the whole negative H, the apparatus described in Fig. 55 may be used. Because the shift between patterns changes from one point to another on H, the different regions of the negative must be isolated by means of an opaque screen with a small aperture. In the focal plane of a lens O, Young's fringes are observed, and their direction and fringe spacing are related to the direction and the magnitude of the shift δl between the two recorded speckle patterns.

The experiment described in Fig. 67 may be done by displacing the diffuse object itself between the two exposures, instead of the photographic plate. It was shown in Section 2.5 that two correlated speckle patterns that are related to each other by a scale factor are recorded on the photographic plate H, and that a relatively large diffuser G may be used. After development, the negative may be observed with the apparatus of Fig. 69.

5.2   A Change of Wavelength between the Two Exposures

Without moving any of the optical elements of the apparatus, two successive exposures are made on the same photographic plate H, as shown in Fig. 70, and only the wavelength is changed. This is the experiment described in Section 2.6. Wavelengths λ and λ' correspond to the first and to the second exposures, respectively. The diffuse object G may be a ground glass. If conditions (2.9) and (2.10) are satisfied, the two speckle patterns recorded successively on H are related by a scale factor equal to $1 - \delta l/l$. These conditions may be satisfied by means of an appropriate choice of wavelengths λ and λ'. Because of the circular symmetry of the system, circular fringes are once again observed when a negative is used as shown in Fig. 69.

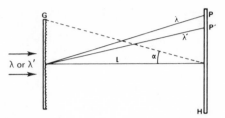

Fig. 70   Recording two speckle patterns that are identical except for a scale factor by means of a change of wavelength.

5.3   A Change of Wavelength and a Displacement of the Photographic Plate between Exposures

Two successive speckle patterns are recorded on a photographic plate as shown in Fig. 71. The first exposure is made with a wavelength λ, and the second exposure is made with a wavelength λ', the holographic plate being moved between the two exposures as in the experiment described in Section 2.7. The photographic plate is at H_1 in the first exposure, and at H_2 in the second exposure, and the displacement δl satisfies relation (2.11). If, in addition, inequalities (2.9) and (2.10) are satisfied, two identical speckle patterns are recorded on the photographic plate. Let us repeat this experiment except that plate H is given two displacements between the two exposures: the previous displacement given by Eq. (2.11) and a lateral displacement. In this case we recorded two identical speckle patterns shifted relative to each other as in the experiment of Burch and Tokarski. Young's fringes are then observed in the spectrum of the negative.

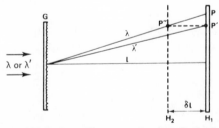

Fig. 71   Recording two identical speckle patterns by changing both the wavelength and the plane of focus.

5.4   Circular Interference Fringes Produced by Two Successive Recordings on the Same Photographic Plate of the Image of a Diffuse Object*

In the foregoing sections the recorded speckle patterns were produced at a finite distance from a diffuse object illuminated with a laser. The same experiment may be repeated in the image of a diffuse object itself. For example, going back to Fig. 19, two speckle patterns may be successively recorded on the same photographic plate located first at P', and then at P''. If condition (1.6) is satisfied, two speckle patterns, which are identical except for a scale factor $1 - \delta l/l$, will be obtained as in the experiment of Section 5.1. When the negative is examined with the system of Fig. 69, the same rings are observed.

If the photographic plate does not move, but the wavelength is changed between the two exposures, the two speckle patterns described in Section 1.9 are recorded. If condition (1.18) is satisfied, the two speckle patterns are practically identical. This may be verified by changing the experiment a bit: between the first recording with wavelength λ and the second recording with wavelength λ', the photographic plate is shifted laterally. This yields the same conditions as the experiment of Burch and Tokarski: Young's fringes are observed in the spectrum of the negative.

5.5   Circular Interference Fringes Obtained with a Single Exposure by Means of an Amplitude Diffuser

In the experiments that follow, the diffuser used changes only the amplitude, not the phase of the incident light. Such an amplitude diffuser may be produced by a technique described in Section 2.2. The diffuser consists of

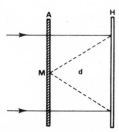

Fig. 72   Recording the speckle patterns produced by A and the unscattered coherent background through A.

* See Archbold and Ennos (1972).

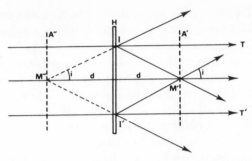

Fig. 73   Reconstruction of the two images A' and A'' of the diffuser A.

a photographic plate on which a speckle pattern is recorded. Of course it is the negative that is used, after development. The experiment corresponds to Fig. 72. A diffuser A is illuminated with a collimated beam of light from a laser. Part of the light goes through A without being scattered, and the rest is scattered by the grains of the speckle such as M. A photographic plate H is put at an arbitrary distance d from A, and in this first experiment a single exposure is made. This experiment is quite similar to the recording of a Gabor hologram, the reference beam being the unscattered beam. After development, when the hologram H is illuminated with a collimated beam of light, it reconstructs two identical and symmetrical diffusers from H, as shown in Fig. 73: a conjugate diffuser A'' and a direct diffuser A'. The conditions are the same as those of Section 3.3. The path difference in a direction i between two parallel rays from two points such as M' and M'' is $\Delta = i^2 d$. This yields rings at infinity that follow relation (3.3), with radii given by Eq. (3.5), with $n = 1$.

5.6   Circular Interference Fringes Obtained with More Than One Exposure When the Photographic Plate Is Axially Translated between Each Exposure*

Unless very small values of d are used in Eq. (5.5), the experiment described in the previous section will yield rings with small angular diameters more or less hidden by the direct beam IT, $I'T'$, as shown in Fig. 73. Indeed if $d = 6$ mm ($\lambda = 0.6$ μm), the angular diameter of the first bright ring is equal to 26′, that is, the order of magnitude of the apparent diameter of the sun.

* See Lifchitz and May (1972).

Experience shows that this is not enough to avoid the inconvenience of the transmitted beam IT, $I'T'$. Therefore with the preceding amplitude diffuser, two successive exposures are made, and the photographic plate is given an axial translation between the two exposures. This yields rings with large apparent diameters if the translation ε of the plate between the two exposures is small enough, and this is easy to do. When the developed hologram H is illuminated with parallel light, four identical and symmetrical diffusers are reconstructed, A_1' and A_2' corresponding to the direct image, and A_1'' and A_2'' corresponding to the conjugate image as shown in Fig. 74. Diffusers A_1' and A_2' produce rings at infinity with angular radii given by Eq. (3.5), which may be written

$$i = \sqrt{\frac{2p\lambda}{\varepsilon}}\qquad(5.4)$$

The diffusers A_1'' and A_2'' yield the same rings. Because the distance between the two groups of diffusers is much greater than ε, the other rings produced by the system are practically in the unscattered beam and therefore invisible.

Rings corresponding to multiple interference may be made by means of a series of successive exposures on the same photographic plate with the same translation ε of the plate between each exposure. If N identical exposures are made, N direct diffusers and N conjugate diffusers, all identical, are reconstructed on either side of the hologram. The rings at infinity then follow the relation

$$I = I_0\left(\frac{\sin(N\pi\varepsilon i^2/2\lambda)}{\sin(\pi\varepsilon i^2/2\lambda)}\right)^2\qquad(5.5)$$

The greater the value of N, the finer the rings, the limit being set by the dynamic range of the photographic emulsion.

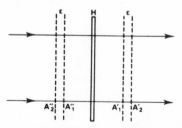

Fig. 74 Same experiment as that described by Fig. 72, but with a small axial translation of H between exposures.

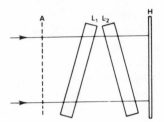

Fig. 75   Recording the speckle produced by *A* in order to obtain hyperbolic or elliptical fringes.

5.7   Hyperbolic or Elliptical Fringes

As shown in Fig. 75, two glass plates L_1 and L_2 with parallel faces are interposed between the amplitude diffuser *A* and the photographic plate *H*. The plane that bisects L_1 and L_2 is parallel to *H*. Two plates are used in order to correct the shift of the rays produced by a single plate; two successive exposures are made on *H*. The first exposure is made under the conditions corresponding to Fig. 29, whereas the second is made after plates L_1 and L_2 have been removed and *H* has been shifted in a direction perpendicular to its plane. Let *d* be the distance between *A* and *H* during the second exposure. The hologram will reconstruct two direct diffusers and two conjugate diffusers. In the group of the two direct diffusers, one is reconstructed with the astigmatism caused by the two glass plates during the recording, whereas the other is reconstructed without aberration and serves as a reference. The interference patterns at infinity produced by those two diffusers are hyperbolic or elliptical fringes depending on the value of *d*. The phenomena are the same for the two conjugate diffusers.

CHAPTER VI

Optical Processing of Images
Modulated by Speckle

6.1   Introduction

When a grating R is illuminated with a parallel beam of light as in Fig. 76, the orders of the grating are observed in the focal plane F of the lens O. If an opaque screen with an aperture small enough that only the direct image of the source located at F goes through is put in the focal plane of the lens, a uniform illumination is observed in the conjugate plane R' of the grating R. There is no image of the grating. If the diameter of the aperture is now increased so that not only the direct image but one spectrum on each side of this image goes through, an image of the grating appears. If the direct image F is blocked out, the grating frequency is doubled in the image R'. In the case of a two-dimensional grating R, the orientation of the bars in image R' may also be changed by means of an appropriate filtering of the

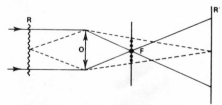

Fig. 76   Filtering the spectrum of a grating.

73

spectra of the grating in the focal plane of lens O. This classical experiment, known as Abbe's experiment, may be considered as one of the first experiments in optical image processing. It has been generalized and applied to arbitrary objects. For example, the blur of an unfocused photograph is due to the presence of low-frequency components. By means of a filter that reduces the light in the immediate vicinity of the source at F (the region of low frequencies), it is possible to improve the quality of the photograph. For an image degraded by noise caused, for example, by granularity, a filter that reduces the light in the regions further away from F (the regions corresponding to the high spatial frequencies) allows the reduction of the granularity. Unfortunately, such a low-pass filter may cut out the high spatial frequencies of the image itself and thus reduce its quality. On the other hand, if the image is degraded with periodic noise, the noise may be suppressed by means of a filter that cuts out the frequencies corresponding to the periodic noise. The high spatial frequencies of the image are not all cut out, and the image is thus less degraded.

These spatial filtering operations are, however, difficult to do in practice, because the spectrum about F is very small. For a lens with an aperture equal to $f/6$, the central spot of the diffraction image has a radius equal to 4 μm. To modify the low spatial frequencies, very small filters, which are difficult to fabricate, are required. In addition, spatial filtering with coherent light always yields images degraded by diffraction parasites produced by dust, scratches, and so on.

To spread out the spectrum in the focal plane of lens O, the image to be processed, for example, a photograph, may be modulated with a diffuser. Of course the diffuser must have a sufficiently fine structure that the quality of the photograph is not modified. Consider the experiment described in Fig. 26. The ground glass G, illuminated by a laser, is used as a light source to illuminate an unexposed high-resolution photographic plate H against which is placed a transparency A that is to be modulated. This is shown in Fig. 77. A speckle pattern is produced by the ground glass G on H, and its intensity may be represented by the function $D(\eta, \zeta)$. Because the slide A is

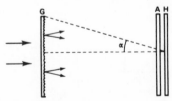

Fig. 77   Recording on H of a photograph A modulated by speckle.

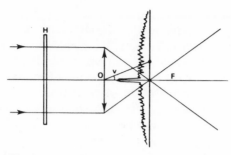

Fig. 78 Spectrum of a negative *H* at the focus of a lens *O*.

immediately before *H*, the intensity on *H* is the product of $D(\eta, \zeta)$ multiplied by the function $A(\eta, \zeta)$, which represents the transmittance of the slide *A*. The product $A(\eta, \zeta)D(\eta, \zeta)$ is therefore recorded. The diameter of the speckle grains will be sufficiently small if *A* is given an appropriate value. If $\alpha = \frac{1}{10}$, the grain size will be of the order of 6 μm (Section 2.3). After development, the negative *H* looks like a normal negative, and its fine structure caused by the speckle may be seen only under a microscope. Under these conditions, the quality of the image may be considered to be unaltered. The spectrum of the negative is then spread out in the focal plane of a lens as shown in Fig. 78. If the lens *O* has a focal length equal to 30 cm, the central spot of the diffraction phenomena produced by the speckle has a diameter of the order of 70 mm. In fact, the spectrum is formed by the convolution $\tilde{A}(u, v) \otimes \tilde{D}(u, v)$, and although it is considerably spread out, this is not sufficient for processing operations to be carried out. This spectrum must be modulated in such a way that the light be localized in certain areas of the spectrum plane, depending on which problem is being solved.

6.2   The Principle of a Technique to Extract the Difference between Two Images*

Consider the following problem: two photographs *A* and *B* have some identical and some nonidentical areas, and the difference $A - B$ is required. Transparencies *A* and *B* might be, for example, two photographs of the same ground area taken at different times, between which some modifications may have taken place. These modifications are to be displayed by removing all the features that have remained unchanged.

* See Debrus *et al.* (1974).

Transparencies *A* and *B* are copied on the same high-resolution photo-graphic plate *H*, as illustrated in Fig. 79. During the first exposure, trans-parency *A* is placed against *H* and is illuminated by means of a speckle pattern from the ground glass *G*. *A* is then replaced by *B*, but before the second exposure, *H* is given a small translation in any direction. Figure 80 repre-sents an area of the negative *H* after development. In areas where trans-parencies *A* and *B* are identical, for each speckle grain recorded with *A*, there corresponds an identical speckle grain recorded with *B*. The distance between two identical grains is equal to the translation given to the plate *H* between the two exposures. On the other hand, in areas where *A* and *B* are different, the intensities of the corresponding points are different. If in a certain area transparency *A* is completely transparent and *B* is opaque, only the speckle grains recorded with *A* will be present. Let the negative

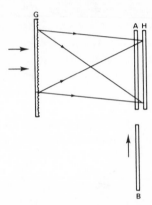

Fig. 79   Successive recordings of two photographs *A* and *B* modulated by the same speckle pattern.

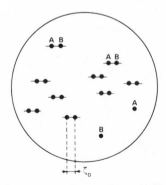

Fig. 80   Structure of the resulting speckle on *H*.

H be illuminated with a collimated beam from a filtered mercury vapor lamp (Fig. 81). The lens O forms an image of H at H'. In areas where A and B are identical, there are groups of two identical grains. These grains diffract light, and two identical grains such as A_1 and B_1 yield Young's fringes in the focal plane F of lens O. This is true of all the areas of A and B that are identical. If an opaque screen with a slit is placed in the focal plane in such a way that the slit coincides with a dark fringe, the light from A_1 and B_1 will not reach the image H'. This is true for every group of two identical grains, and therefore the light from all the identical areas of A and B will disappear from image H'. Consider an area where transparencies A and B are different. The two corresponding grains are different and do not diffract the same luminous intensities. They yield Young's fringes with nonzero minima. Some of the light goes through and reaches H'. Only areas of A and B which are different are seen in H': the difference $A - B$ is observed.

6.3   The Light Distribution in the Image Plane

The technique just described may be explained by means of a very simple mathematical development. In the first exposure, transparency A is placed against the unexposed high-resolution plate H, and the product $A(\eta, \zeta)D(\eta, \zeta)$ is recorded, where $A(\eta, \zeta)$ represents the absorption of transparency A, and $D(\eta, \zeta)$ is the speckle intensity. Transparency A is replaced by transparency B in such a way that the regions of B that are identical to certain regions of A are located exactly in the same places. Photographic plate H is then given a small translation in any direction, say in a direction parallel to the ζ axis. Let this translation be ζ_0. The second exposure is made under those conditions. The intensity $B(\eta, \zeta - \zeta_0)D(\eta, \zeta - \zeta_0)$ is therefore recorded. Because a translation is equivalent to convolution with a delta function (Sections 1.6 and 4.2), the recordings may be expressed as $[B(\eta, \zeta)D(\eta, \zeta)] \otimes \delta(\eta, \zeta - \zeta_0)$. In sum, the recorded intensity on H may be expressed in the abbreviated form

$$(AD) \otimes \delta(\eta, \zeta) + (BD) \otimes \delta(\eta, \zeta - \zeta_0) \tag{6.1}$$

Now set $A - B = C$:

$$(AD) \otimes [\delta(\eta, \zeta) + \delta(\eta, \zeta - \zeta_0)] - (CD) \otimes \delta(\eta, \zeta - \zeta_0) \tag{6.2}$$

which may be expressed in the symmetrical form

$$(AD) \otimes \left[\delta\left(\eta, \zeta + \frac{\zeta_0}{2}\right) + \delta\left(\eta, \zeta - \frac{\zeta_0}{2}\right)\right] - (CD) \otimes \delta\left(\eta, \zeta - \frac{\zeta_0}{2}\right) \tag{6.3}$$

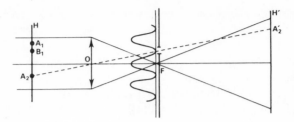

Fig. 81 Detection at H' of the difference between two images recorded on H.

After development, the amplitude t transmitted by the negative is equal to (Section 4.2)

$$t = a - b\left\{(AD) \otimes \left[\delta\left(\eta, \zeta + \frac{\zeta_0}{2}\right) + \delta\left(\eta, \zeta - \frac{\zeta_0}{2}\right)\right]\right.$$

$$\left. - (CD) \otimes \delta\left(\eta, \zeta - \frac{\zeta_0}{2}\right)\right\} \tag{6.4}$$

The negative H is observed as shown in Fig. 81. It is illuminated with a collimated beam of light, and the lens O yields an image at H'. In the focal plane of lens O, the spectrum of the negative is given by the Fourier transform of the amplitude transmittance t. The results of Section 4.2 show that

$$\tilde{t}(u, v) = a\delta(u, v) - b\left\{2\left[\tilde{A}(u, v) \otimes \tilde{D}(u, v)\right]\right.$$

$$\left. - \cos\left(\frac{\pi v \zeta_0}{\lambda}\right) - \left[\tilde{C}(u, v) \otimes \tilde{D}(u, v)\right]\exp(-j\pi v \zeta_0/\lambda)\right\} \tag{6.5}$$

The first term $a\delta(u, v)$ on the right-hand side of the equation represents the image of the point source that illuminates the system, localized at F. This image may be neglected, because of its small dimensions. In the second term, except for a constant b, the spectrum $\tilde{A}\tilde{D}$ is modulated by the factor $\cos(\pi v \zeta_0/\lambda)$, which represents Young's fringes. The angular distance between two consecutive bright or dark fringes is equal to λ/ζ_0. Let an opaque screen with a slit be placed in the focal plane of lens O in such a way that the slit coincides with a minimum of the Young's fringes. The term

$$(\tilde{A} \otimes \tilde{D})\cos\left(\frac{\pi v \zeta_0}{\lambda}\right) \tag{6.6}$$

is blocked out, and except for a constant factor, the screen allows only the term

$$(\tilde{C} \otimes \tilde{D}) \exp(-j\pi v\zeta_0/\lambda) \tag{6.7}$$

to go through to image H', where the amplitude $C \times D$ is incident. Only the difference $C = A - B$ is visible, this difference being modulated by the speckle D. Because this speckle is very fine, it does not affect the quality of image C.

6.4   Improving the Profile of the Fringes and the Quality of the Images

The curve labeled (1) in Fig. 82 shows the profile of the amplitude of the fringes that are observed in the focal plane of lens O in Fig. 81. From Fig. 82, the slope of curve (1) is equal to unity near the zero value. In order to let through the least amount of light from the identical regions of A and B, a narrow slit must be used, and this is not favorable to the quality of the filtered image. A much wider slit may be used if the results of Section 4.3 are taken into account. By taking an odd number of successive exposures, with exposure times proportional to the binomial coefficients, narrow bright fringes separated by wide intervals of zero minima may be obtained. Because of the limited dynamic range of the photographic emulsion, the number of

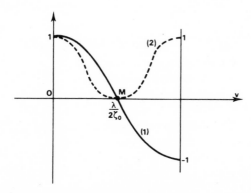

Fig. 82   Structure of the fringes in the spectrum for two exposures [curve (1)] and for three exposures [curve (2)].

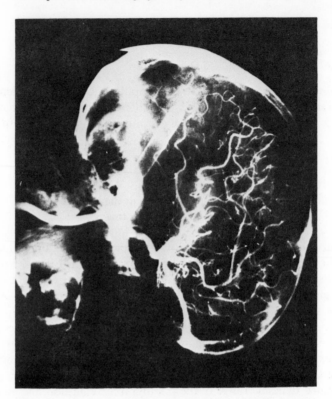

Fig. 83   X ray of a human skull before injection of a liquid in the arteries.

exposure is limited, but three exposures already yield good results. The recordings are done as follows:

images	A	B	A
translation	$-\zeta_0$	0	$+\zeta_0$
exposure time	$\frac{1}{2}$	1	$\frac{1}{2}$

The recorded intensity is equal to

$$(AD) \otimes [\tfrac{1}{2}\delta(\eta, \zeta - \zeta_0)] + (BD) \otimes \delta(\eta, \zeta)$$
$$+ (AD) \otimes [\tfrac{1}{2}\delta(\eta, \zeta + \zeta_0)] \qquad (6.8)$$

which may be written

$$(AD) \otimes [\tfrac{1}{2}\delta(\eta, \zeta - \zeta_0) + \delta(\eta, \zeta) + \tfrac{1}{2}\delta(\eta, \zeta - \zeta_0)] - [(A - B)D] \otimes \delta(\eta, \zeta)$$
$$(6.9)$$

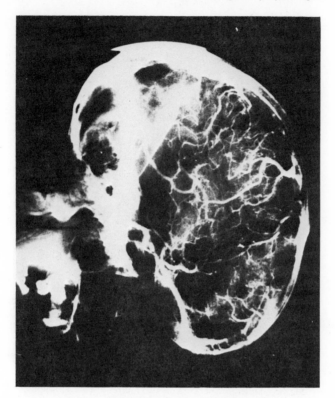

Fig. 84   X ray taken after injecting the liquid.

After expressing the transmitted amplitude t of the negative after development as in the preceding section, the spectrum of the negative is equal to

$$\tilde{t}(u, v) = a\delta(u, v) - b\left[2(\tilde{A} \otimes \tilde{D})\cos^2\left(\frac{\pi v \zeta_0}{\lambda}\right) - \tilde{C} \otimes \tilde{D}\right] \qquad (6.10)$$

where $C = A - B$. The identical regions of A and B are now amplitude modulated by the factor $\cos^2(\pi v \zeta_0/\lambda)$, represented by the curve labeled (2) in Fig. 82. Because of the zero minimum at M, it is possible to use a much wider slit and to improve considerably the quality of the images. In order to record the three images with only two exposures, the birefringent system described in Section 4.4 may be used. This apparatus automatically positions the speckles, and it is not necessary to translate the plate between the two exposures. Figures 83–85 show an example applied to the processing of x rays of a human skull. In Fig. 84, a liquid has been injected into the arteries of the

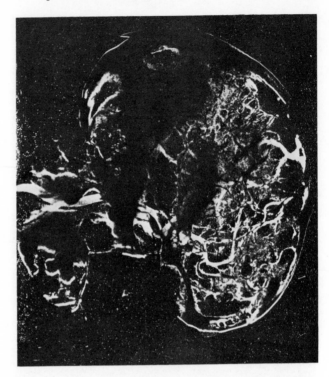

Fig. 85   Difference between the two images in Figs. 83 and 84.

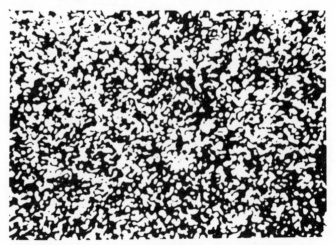

Fig. 86   Structure of the speckle as seen under a microscope.

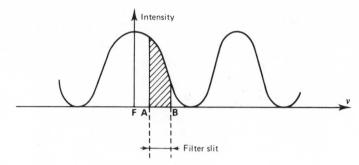

Fig. 87   Influence of the width of the slit filter.

brain, and the propagation of the liquid may be followed by taking the difference with Fig. 83, which was taken before the injection. Figure 85 shows the difference between the two photographs of Figs. 83 and 84. The light contour around the image of Fig. 85 shows that the two x rays 83 and 84 have not been taken exactly under the same conditions. However, this contour facilitates the location of the features relative to the skull. Figure 86 is a photograph of the speckle as seen under a microscope with a high magnification.

It has been shown that by giving the exposure times values proportional to the binomial coefficients, the image quality may be improved considerably. However, in the preceding considerations we have not taken into account the width of the slit in the filter. For a given slit, the profile of the fringes that yields the least possible light in the image must be found. This is equivalent to studying the variation of the hatched area on Fig. 87 when the segment *AB*, which represents the width of the slit, is displaced along the axis *ov*. The profile of the fringes that yields an absolute minimum is required, and this determines the exposure times (Marom and Kasher, 1977).

6.5   Image Coding and Decoding

The preceding technique for extracting differences may also be applied to the decoding of coded images. Let *A* be a transparency that represents an arbitrary object that must be coded then decoded. Two successive exposures are made by contact on an unexposed photographic plate *H*. The first is made with *A* and the second with a transparency *B*, which degrades *A* enough that it is not recognizable. For example, transparency *B* might be a random distribution of points, of lines, or arbitrary curves, and so on.

After development, photograph H is the transmitted message that must be decoded by the user. The latter has the transparency B, which is the "key" for decoding. The two photographs $H = A + B$ and B are processed with the technique of Section 6.2, that is, the difference $A = H - B$ is obtained. It is of course impossible to extract image A without the "key" B.

6.6   Image Multiplexing by Superposition of Laterally Shifted Speckle Patterns*

Consider again Fig. 77. Transparency A is recorded on the photographic plate H. A number of exposures are made with exposure times proportional to the binomial coefficients, and with the same translation between each exposure. After development, the negative H is observed as shown in Fig. 81. Because a number of exposures have been made, the fringes observed in the focal plane of lens O are narrow. For simplification, they have been represented by straight lines such as $A_1, A_2, \ldots, A_{-1}, A_{-2}, \ldots$ on Fig. 88. The experiment is then repeated on a new photographic plate H. Having thus recorded the transparency A on H, the same is done with another transparency B, but the translation between exposures is just a little bit smaller than that given to H between the exposures done with A. A and B have thus been recorded on the same photographic plate H. After development, the spectrum is observed as in Fig. 81. This spectrum is represented in Fig. 89. The fringes due to A and B do not coincide, because the translations between

Fig. 88   Spectrum of an image recorded a large number of times when the same small translation is given to the photographic plate between each exposure.

Fig. 89   Recording of two images with different translations.

* See Grover (1972a).

exposures are different. It is easy to see that if the slit filter is placed on a fringe such as $A_1, A_2, \ldots, A_{-1}, A_{-2}, \ldots$, only the image of A will appear in the image plane H' of Fig. 81. If the slit filter is on a fringe such as B_1, $B_2, \ldots, B_{-1}, B_{-2}, \ldots$, it is the image of B that appears at H'. Of course, the number of photographs that may be recorded on H is not unlimited. The dynamic range of the emulsion does not allow more than five or six images to be recorded without degradation of the signal-to-noise ratio.

6.7 Image Multiplexing with Oriented Speckle Patterns*

Figures 90–92 show three methods to superimpose images on the same photographic plate H. In the case of Fig. 90, one of the images A to be recorded is placed near a ground glass G illuminated with a laser. A lens O forms an image of A on a photographic plate. A diaphragm in the shape of a slit is put on lens O, and this causes the speckle grains to be small diffraction spots elongated in a direction perpendicular to the direction of the slit. A second photograph B is recorded, with a different orientation of the slit on lens O. After development, the photographic plate is placed on H as shown in Fig. 81. In the spectrum corresponding to A, the diffracted light is obviously not oriented in the same direction as in the spectrum corre-

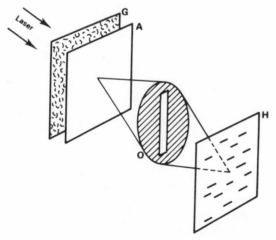

Fig. 90 Image multiplexing on H by changing the orientation of a slit.

* See Kopf (1974).

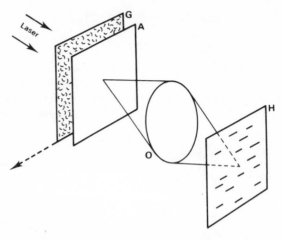

Fig. 91   Image multiplexing on *H* by translating the ground glass *G* in different directions.

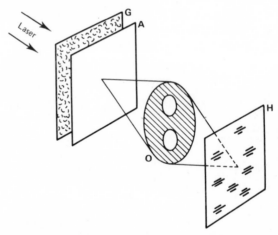

Fig. 92   Image multiplexing on *H* by rotation of a diaphragm with two apertures.

sponding to *B*. If the slit filter at *F* is oriented properly on Fig. 81, images *A* or *B* may be made to appear.

A similar procedure is shown in Fig. 91, but the elongation of the speckle grains is obtained simply by giving the ground glass a uniform translation during the exposure. It suffices to choose a different translation for each recorded image. After development, the successive observation of the different images may be done by changing the orientation of the slit filter as before.

Finally, in Fig. 92, the images are multiplexed by superimposing two identical apertures to the lens. Each speckle is similar to a diffraction spot modulated by Young's fringes. The orientation of the fringes depends on the orientation of the two apertures. This orientation is changed for each recorded image. After development, the successive observation of the images with the apparatus of Fig. 81 is done as before, by changing the orientation of the slit filter. In every case, the dynamic range of the photographic plate limits the number of images to five or six.

CHAPTER VII

The Study of Displacements
and Deformations of Diffuse Objects
by Means of Speckle Photography

7.1 The Study of Lateral Displacements of a Diffuse Object
When the Displacement is Greater Than the Diameter
of a Speckle Grain*

The object A under consideration is illuminated with a laser and the image A' of A formed by the lens O is observed as shown in Fig. 93. For simplicity, suppose that the object A is formed of two parts A_1 and A_2. Part A_1 remains stationary during the experiment, whereas part A_2 moves. This does not reduce in any way the generality of the following results. We consider the case of a lateral displacement: part A_2 therefore moves in a

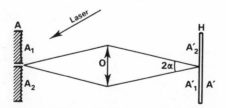

Fig. 93   Study of the lateral displacements of an object A.

* See Leendertz and Butlers (1971).

89

direction perpendicular to the optical axis of lens O. The image A' is recorded
on a photographic plate H by means of two successive exposures, one before
the displacement of A_2, and the other after. Between the two exposures, the
photographic plate H is given a small translation ζ_0 in its plane and in any
arbitrary direction. Within the area A_1', two identical speckle patterns shifted
relative to each other by the distance ζ_0 are recorded. Because part A_2 has
been shifted between the two exposures, area A_2' of the photograph is different. If the displacement of A_2 consists of a simple translation, the photograph will contain in A_2' two identical speckle patterns shifted relative to
each other by a distance not equal to ζ_0. After development, the image of
the negative H' is observed with the classical experiment of Fig. 94, which is
similar to Fig. 81. The negative H is illuminated with collimated light by
means of an ordinary monochromatic light source, for example, a filtered
mercury vapor lamp. The lens L forms an image of H at H'. In the focal
plane F of lens L, area A_1' yields straight parallel fringes with a spacing
equal to λ/ζ_o (see Section 4.2). Area A_2' yields fringes whose orientation will
depend on the direction of the translation and with a spacing different
from λ/ζ_0. If a diaphragm with a slit aperture is placed in the focal plane F
in such a way that the slit coincides with an absolute minimum of the fringes
produced by A_1', the image A_1'' of A_1' will disappear. Because the fringes
produced by A_2' do not coincide with the aforementioned fringes, some light
goes through and reaches the image A_2'' of A_2'. An image of part A_2 of object
A, which has been displaced, is observed at H' with an excellent contrast.
The areas that did not move have disappeared.

We have considered a simple example where one part of the object remains
stationary, whereas another part is translated. Of course the result is general,
and the method detects all the areas of the object that are translated or deformed. When two exposures are made, the slit filter must be narrow, which
reduces the quality of the image at H'. We have previously seen (Sections 4.3,
4.4, and 6.4) that the quality may be improved by recording three images of
the object with exposure times proportional to 1, 2, 1, either with three ex-

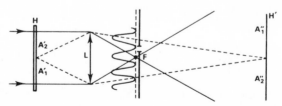

Fig. 94 Filtering the spectrum of the negative H obtained in the experiment of Fig. 93.

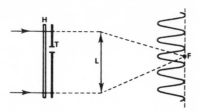

Fig. 95   Observation of the spectrum of negative H when a diaphragm with an aperture T is placed against the negative.

posures, or with two exposures by means of a birefringent system (Section 4.4). For a quantitative study, we may proceed point by point as shown in Fig. 95 (Archbold and Ennos, 1972). Instead of observing the image of negative H, the fringes are examined in focal plane F of lens L while the diaphragm with an aperture T is placed against the negative H. The orientation and the fringe spacing changes with the position of aperture T. The direction and magnitude of the displacements of the various areas of the object may therefore be measured.

It should be noted that this method applies only if the displacement of the object in plane A' is greater than the diameter of a speckle grain. Indeed, suppose that we had made only one exposure in the experiment of Fig. 93. Only one speckle pattern is recorded, and its spectrum is observed as shown in Fig. 94. The speckle is composed of small grains. the smallest having a diameter approximately equal to the size of the diffraction pattern of lens O used to record the speckle. Therefore the smallest speckle grains on H have diameters of the order of $\varepsilon = \lambda/\alpha$. In the experiment of Fig. 94, most of the light is contained in a cone with a half-angle equal to $\alpha = \lambda/\varepsilon$. In order to see fringes in the case of a double exposure, the fringe spacing must not be too great, that is, the first absolute minimum of the fringes must be inside the cone of diffracted light. If the displacement of the object measured in the image plane A', that is, on H, is equal to ζ_1, the first absolute minimum will be found in a direction making an angle of λ/ζ_1 with the optical axis. Fringes will be observed if this angle is smaller than the half-angle $\alpha = \lambda/\varepsilon$ of the cone of diffracted light, that is, if $\zeta_1 > \varepsilon$. The displacement in the image plane must therefore be greater than the diameter of a speckle grain. On the other hand, the displacement must not be too great either, because the fringe spacing becomes very small, and measurements are hindered by the direct image of the source, which is located at F in Fig. 94. The limits within which the experiment may be made are easily determined for each type of configuration.

7.2   The Study of Lateral Displacements of a Diffuse Object When the Displacement Is Smaller Than the Diameter of a Speckle Grain*

Consider now the case where the displacement measured in the plane of the image A' of object A (Fig. 93) is smaller than the diameter of a speckle grain in plane A'.

Let a diaphragm with two slit apertures T_1 and T_2 be located in front of lens O, which forms the image A' of a diffuse object A, as shown in Fig. 96. In the image A' of A, due to the light that passes through T_1, there is a speckle pattern that is uncorrelated with the speckle pattern due to the light that passes through T_2. The speckle observed in the image A' results from the interference between those two speckle patterns. This is a case similar to that of Section 3.5. The two slits produce two different speckle patterns just like the two ground glasses G_1 and G_2 of Fig. 49. The speckle grains observed on A' display straight and parallel interference fringes with a fringe spacing equal to $\lambda D/\overline{T_1 T_2}$, where D is the distance between O and the image plane A'. As seen in Section 3.5, there is no relationship between the positions of the fringes as we pass from one grain to another in the speckle pattern. If each slit has a width ζ, the diameter of the speckle grains is of the order of $\lambda D/\zeta$. Because $\overline{T_1 T_2} \gg \zeta$, the fringe spacing is obviously much smaller than the grain diameter. For simplicity, suppose that the object A is translated in a direction perpendicular to the direction of the slits (the slits are perpendicular to the plane of Fig. 96). Two successive exposures are made on the same photographic plate, one before the translation and the other after. The translation of the object A produces a simple translation of the speckle at A'. All the grains together with their fringes are translated. Suppose that the translation of the object A is such that in the second exposure the fringes are shifted by a distance equal to half the fringe spacing. The maxima corresponding to the first exposure are then superimposed to the minima corre-

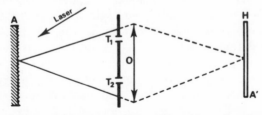

Fig. 96   Study of the lateral displacement of a diffuse object, when the speckle grains are modulated by interference fringes.

* See Duffy (1972).

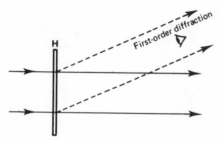

Fig. 97   Observation of the negative from one of the diffracted orders.

sponding to the second exposure: the fringes disappear in all the grains. This will be the case each time that the displacement of the object A produces in A' a shift equal to an odd multiple of $\lambda D/2\overline{T_1 T_2}$. The shift must remain small relative to the speckle grain diameter; otherwise the fringes will not completely disappear.

Let us now assume that the object A is deformed unequally, and two successive exposures are made on the same photographic plate. After development, the plate is illuminated with collimated light, and the plate is observed from one of the diffracted orders, outside of the directly transmitted beam, as shown in Fig. 97. The areas that have been shifted by a distance equal to an odd multiple of $\lambda D/2\overline{T_1 T_2}$ will appear dark. The areas that have been shifted by a distance equal to an even multiple of $\lambda D/\overline{T_1 T_2}$ will appear bright. The contrast of these fringes is maximum, that is, it is equal to unity. This setup is very simple, but the use of the slits reduces the brightness.

It is possible to use an auxiliary diffuser in order to improve the detection of very small lateral displacements. The experiment of Fig. 93 is carried out with a ground glass against the photographic plate H; the contrast of the fringes in the spectrum of the negative decreases when the displacement of the object increases, and it becomes equal to zero when the displacement reaches a value of the order of the diameter of a speckle grain.

7.3   The Study of Lateral Displacements of a Diffuse Object Illuminated with Two Beams When the Displacement Is Smaller Than the Diameter of a Speckle Grain*

Let a diffuse object A be illuminated with two coherent laser beams inclined by the same angle θ to the normal of object A as shown in Fig. 98.

* See Leendertz (1970).

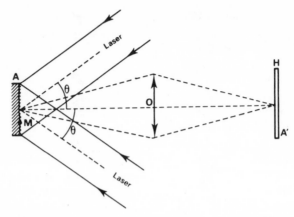

Fig. 98   Study of the displacement of a diffuse object *A*, illuminated by two symmetrical coherent beams.

An image of *A* is formed at *A'* by means of a lens *O*. The speckle observed at *A'* results from the interference of two uncorrelated speckle patterns produced by the two beams. In Fig. 98, the direction of the rays in the two beams is parallel to the plane of the figure, and the plane diffuse object *A* is perpendicular to the plane of the figure and to the optical axis of lens *O*. Let us consider a displacement of object *A* in a direction perpendicular to the optical axis of lens *O*. Let the path difference between the two rays from the two beams incident at any point *M* of *A* be equal to δ. If the object *A* is shifted by ζ, the optical path increases by $\zeta \sin \theta$ for one of the two rays incident at *M*, and it decreases by the same quantity for the other. The path difference becomes $\delta \pm 2\zeta \sin \theta$. If this path difference is equal to an integral multiple of the wavelength λ, everything happens as if the path difference at *M* between the two rays from the two beams had its initial value. In this case the speckle pattern at *A'* becomes identical to the speckle pattern observed before the displacement of the object *A*.

The phenomenon may be observed in the following manner: the speckle is recorded on photographic plate *H* before the displacement. After development, the negative is put exactly in the same place in the apparatus used for the recording, and the surface of the negative is observed. The speckle produced by the object *A* is seen through its complement, recorded on the negative. The negative *H* appears uniform and dark. Let the object now be displaced as previously described. For an arbitrary displacement, the speckle at *A'* is uncorrelated with the initial speckle: the dark areas of the negative do not coincide any more with the light areas of the incident speckle. The negative looks brighter because the transmitted light increases. If the object *A* is locally deformed, luminous zones appear, which correspond to the deformed areas. These luminous zones, called "speckle correlation fringes" have a relatively low contrast.

The technique of double exposure, which consists of taking a photograph of the object *A* on the same plate *H* before and after the displacement (Archbold *et al.*, 1970), may also be used here. Because of the nonlinearity of the photographic emulsion, density variations that display the decorrelation fringes may be seen in transmission. It is preferable to translate the photographic plate slightly between the two exposures. The undeformed areas yield a spectrum that is modulated by fringes with a contrast of unity, which may be eliminated by means of filters with the apparatus shown in Fig. 94. The deformed areas then appear with a very good contrast.

This two-beam method is limited to displacements smaller than the diameter of the speckle grains. The visibility of the decorrelation fringes decreases as the displacement is increased, and it becomes practically zero when the displacement is of the order of magnitude of the diameter of the speckle grains. Very precise real-time measurements may be made by illuminating the diffuse object with two interfering beams combined with a small detector in the image plane (Joyeux and Lowenthal, 1971).

7.4   The Study of the Lateral Displacement of a Diffuse Object with a Diffuse Reference Surface*

This is the experiment of Section 3.5. For simplicity, consider two diffuse objects consisting of two ground glasses G_1 and G_2 as shown in Fig. 99. The

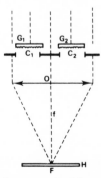

Fig. 99   Study of the lateral displacement of a diffuse object by means of moiré fringes.

* See May and Françon (1976).

two ground glasses are illuminated with the same laser, and the observation is made in the focal plane of a lens O. It is required to study the lateral displacement of object G_1, for example, the other object G_2 remaining stationary. In this experiment an image of the object is not observed and the method therefore only applies to a rigid diffuse object that moves without deformation. We have seen that the two ground glasses G_1 and G_2 yield in the focal plane F two uncorrelated speckle patterns and that because of the interference between those two speckle patterns, the resulting speckle grains are modulated with interference fringes. Let this speckle be recorded on a photographic plate H before the displacement of object G_1. Let a second exposure be made on H after the lateral displacement of G_1. The shifting of the object G_1 has not changed the structure of the speckle, but only the fringe spacing in each grain, and the change is the same for each grain. If the two diffusers are identical, the speckle grains on H have a diameter approximately equal to the size of the diffraction pattern produced by an aperture with the same dimensions as G_1 or G_2. After development, there are two fringe systems with different periods on plate H and on all the grains of the speckle pattern. There is on H a moiré pattern formed of straight parallel and equidistant fringes. The distance between two consecutive fringes is equal to $\lambda f/\zeta_0$, where ζ_0 is the lateral translation given to G_1 and f is the focal length of lens O. These fringes may be observed with a maximum contrast equal to unity if the observer is careful to locate himself outside of the direct beam transmitted by the plate. The accuracy may be easily determined from the characteristics of the apparatus.

7.5   The Study of the Axial Displacement of a Diffuse Object with a Diffuse Reference Surface*

The techniques described in Sections 7.1–7.3 are not very sensitive to axial displacements of diffuse objects. The following experiments may be used to measure small axial displacements. Consider a Michelson interferometer in which the two mirrors are replaced by plane diffuse surfaces M_1 and M_2 as shown in Fig. 100. A lens O forms at H an image of the surface M_1 whose displacement in a direction perpendicular to its plane must be determined. In plane H is observed a speckle pattern resulting from the interference between the two speckle patterns produced by M_1 and M_2. The interferometer is adjusted in such a way that the path difference of the rays reflected by the average surfaces of the two diffusers M_1 and M_2 is

* See Leendertz (1970).

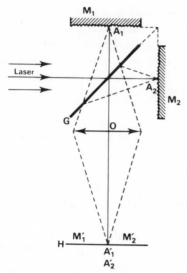

Fig. 100   Observation of the axial displacement of a diffuse object M_1 by means of a diffuse reference surface M_2.

small. The resulting speckle at H is not modulated by interference fringes. It has the aspect of ordinary speckle. Let surface M_1 be given an axial translation small enough that the defocusing that results in the image at H be negligible. The translation is, for example, smaller than $\lambda/2$. Because of the variation in the path difference, the speckle pattern at H changes completely. But when the translation reaches $\lambda/2$, the change in the path difference being equal to λ, the intensities of the speckle grains become equal to their initial values, and the resulting speckle is identical to the initial speckle before the displacement of M_1. The speckle pattern at H becomes identical to itself each time the path difference changes by an integral multiple of the wavelength. But when the displacement becomes greater than a certain value, the defocusing of the image of M_1 on H modifies the speckle structure produced by M_1. The resulting speckle also changes and there are no positions for which the speckle becomes identical to itself. The allowable displacement depends on the aperture α of the lens O that forms the image. Equation (1.6) may be used, the available displacement δl measured in the image space in the vicinity of H being smaller or equal to $\lambda/2\alpha^2$.

Let the speckle pattern before the displacement be recorded on the photographic plate H. After development, let H be put exactly in the same location in the apparatus. As in the experiment of Section 7.3, the dark areas of the negative coincide with the bright grains of the speckle pattern, and the surface of the negative appears uniform and dark. When surface M_1 is given an axial translation, because the intensities of the real speckle grains change, the bright grains do not coincide any more with the dark areas of negative H, and the transmitted light increases. When the translation is equal to $\lambda/2$, the initial state has been reached, and the light transmitted by H is then a minimum.

Let M_1 be located so that the light transmission of H is a minimum. If M_1 is then deformed, the deformed areas will immediately appear for reasons previously given. Decorrelation fringes similar to those described in Section 7.3 will cause variations of the transmitted light. A technique of double exposure may also be used here; the first exposure is made before the deformation, and the second one after. The plate is given a small translation between the two exposures, so that the undeformed areas yield a spectrum modulated with fringes having a contrast of unity. These fringes are then eliminated by means of a filter with the apparatus of Fig. 94. The deformed areas then appear with a good contrast.

7.6   The Speckle Observed in the Focal Plane of Lens O*

Let us reconsider the experiment of Fig. 100, where the speckle is observed in the focal plane of lens O instead of in the image plane of the diffuse object M_1. It was shown in Section 3.6 that the resulting speckle pattern is modulated by circular fringes. An axial displacement of M_1 or M_2 does not de-

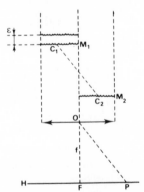

Fig. 101   Study of the axial displacement of a diffuse object by means of moiré fringes.

* See May and Françon (1976).

correlate the speckle, but changes only the radius of the circular fringes. If two successive exposures are made on H, the first before the displacement of M_1 and the second after, a moiré pattern consisting of circular fringes centered on F will be obtained. These fringes allow the measurement of the displacement of M_1. Again as in Section 7.4, this technique may be applied only to a rigid object that is displaced without deformation, because the image of the diffuse object is not observed.

In order to see clearly the moiré fringes, it is preferable to give the two surfaces M_1 and M_2 a lateral shift relative to each other before the exposures. This is shown if Fig. 101, where the surfaces M_1 and M_2 are transparent diffuse objects. The speckle grains at H are modulated by segments of circular fringes with centers at P. The speckle is recorded and M_1 is given an axial translation equal to ε. Then the second exposure is made. After development, the negative H is observed from one of the diffracted orders outside of the directly transmitted beam. The radius $f\sqrt{2\lambda/\varepsilon}$ of the first ring of the moiré yields the axial displacement ε.

7.7 The Use of an Auxiliary Speckle Pattern to Illuminate a Diffuse Surface

When a diffuse object illuminated with a laser is given an axial displacement parallel to the direction of observation, the speckle pattern observed in the image plane changes very little. This is due to the fact that a diffraction pattern does not change very much as a function of the defect of focus (see

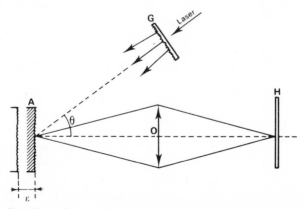

Fig. 102 Detection of the axial displacement of a diffuse object A illuminated with an auxiliary speckle pattern.

Section 1.8). In order to improve the detection of an axial displacement, the experiment of Fig. 37 may be used (see Section 2.8). The diffuse object A of Fig. 37 is illuminated with a speckle pattern produced by the ground glass G, which is illuminated with a laser beam (May and Françon, 1976). This type of illumination was first used to study the vibrations of a diffuse object, as we shall see later (Eliasson and Mottier, 1971). See Fig. 102.

Let Φ be the diameter of the diffraction pattern of lens O in plane A. If the diameter of the speckle grains produced by G on A is of the order of magnitude of Φ, an axial displacement ε of object A will produce a lateral shift of the speckle pattern produced by G, equal to $\varepsilon \tan \theta$, where θ is the angle of incidence of the incident beam. If an appropriate value for angle θ is chosen, a very small axial displacement ε will produce a decorrelation of the speckle pattern at H. If two exposures are made on the photographic plate before and after the displacement, the displacements of the various areas of the object may be detected by filtering the negative as previously described.

7.8   The Study of the Rotation of a Diffuse Surface*

It was shown in Section 4.8 that the rotation of a diffuse reflecting object produces a rotation of the speckle pattern at a finite distance, as if the diffuse surface were a mirror. Near normal incidence, as shown in Fig. 103, a small rotation θ of the diffuse surface A about an axis located in its plane causes the speckle pattern to rotate by an angle 2θ. On a screen E located at a dis-

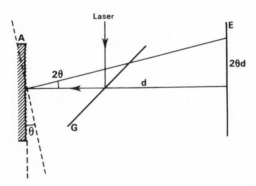

Fig. 103   Study of the rotation θ of a diffuse object A.

* See Archbold and Ennos (1972).

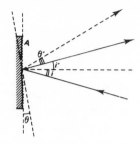

Fig. 104   Relation between the rotation θ of the object and the rotation θ' of the speckle.

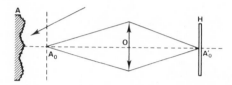

Fig. 105   Study of the rotation of a diffuse object A by forming a defocused image on H.

tance d, the speckle is shifted by $2\theta d$. Let i be the angle of incidence of the illuminating beam, and i' the angle of observation as shown in Fig. 104; the rotation θ' of the speckle pattern caused by a rotation θ of the diffuse surface A is equal to

$$\theta' = \left(1 + \frac{\cos i}{\cos i'}\right)\theta \tag{7.1}$$

This property may be used to study the local deformations of a diffuse surface when these deformations cause changes in the orientation of the surface. The apparatus used could be that of Fig. 105. The diffuse surface A is illuminated with a laser, and an image of plane A_0, which is not identical to A, is formed on a photographic plate H by means of a lens O. If the plate were in the image plane of the diffuse object A, it would obviously be impossible to observe the deformations. If the distance between A_0 and A is equal to d, a rotation θ of A produces, near normal incidence, a rotation $2\theta d$ in plane A_0. The technique of double exposure may be used in recording the speckle on photographic plate H. Because of the defect of focus, the phenomenon is integrated on a surface element proportional to AA_0. In the case of a rigid body rotating without deformation, the speckle may be observed at infinity, that is, in the focal plane of a lens (Tiziani, 1971). Near the normal to a diffuse object, a rotation θ of the object produces a shift

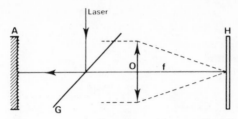

Fig. 106   Study of the rotation of a diffuse object A by recording the speckle in the focal plane of a lens O.

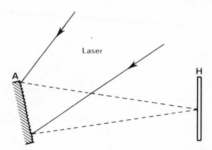

Fig. 107   Study of the rotation of a diffuse object A by recording the speckle in a plane where the image would form if A were a mirror.

$2\theta f$ of the speckle in the focal plane of lens O, whose focal lens is equal to f, as shown in Fig. 106. In the case of Fig. 103, a lateral shift of the object also produces a translation of the speckle in plane H, which may be undesirable. This does not happen in the experiment corresponding to Fig. 106, but this experiment is restricted to rigid objects. Another method is that shown in Fig. 107 (Gregory, 1976). The diffuse surface A is illuminated with a convergent beam, and the photographic plate H is put in the plane where the image of the source would be if the diffuse surface A were a mirror. We have seen in Section 2.3 that in this case a lateral shift of the object A will produce no change in the speckle pattern at H. On the other hand, a rotation θ of object A results in a translation $2\theta d$ of the speckle in plane H at a distance d from A. Here too the phenomenon is integrated on the whole surface of the diffuse object, which must be rigid.

7.9   The Study of the Vibrations of a Diffuse Object

The methods described in the preceding sections to study the rotation of a diffuse object may be applied to the study of the vibrations of diffuse objects. It is possible for instance to use the apparatus of Figs. 103, 105 (Archbold and Ennos, 1972), or 106 (Tiziani, 1971). In every case, if the phenomenon is to be studied by photography, the lateral displacements of the recorded speckle patterns follow a relation that is determined by the kind of vibration under consideration. Consider Fig. 103 and let A vibrate laterally in its plane. A photographic plate located at E records the speckle pattern produced by the diffuse surface A. Let us consider the simple example of a harmonic vibration $f(t) = a \sin \omega t$ in a given direction. It was seen in Section 4.2 that if a double exposure is made, the illumination of the photographic plate is given by Eq. (4.3):

$$D(\eta, \zeta) \otimes [\delta(\eta, \zeta) + \delta(\eta, \zeta - \zeta_0)] \tag{7.2}$$

where ζ_0 is the translation of the photographic plate (or of the speckle) between the two exposures. In the case of one continuous exposure of duration τ during which the speckle is shifted laterally by $\zeta_0 = f(t)$ in a direction parallel to the axis ζ, the preceding expression yields

$$D(\eta, \zeta) \otimes \int_{-\tau/2}^{+\tau/2} \delta[\eta, \zeta - f(t)] \, dt \tag{7.3}$$

Now the amplitude transmitted by the photographic plate after development is proportional to the intensity of the illumination given by the preceding expression. If the negative is illuminated with collimated light, the amplitude spectrum of the negative observed in the focal plane of a lens will be proportional to the Fourier transform of Eq. (7.3). The amplitude in the focal plane of the lens is therefore proportional to

$$\tilde{D}(u, v) \int_{-\tau/2}^{+\tau/2} e^{jkvf(t)} \, dt \tag{7.4}$$

Except for the factor $\tilde{D}(u, v)$, the amplitude in the focal plane is proportional to the integral

$$\int_{-\tau/2}^{+\tau/2} \exp(jkva \sin \omega t) \tag{7.5}$$

This integral may be expanded in terms of Bessel functions, which yields

$$J_0(kva) \int_{-\tau/2}^{+\tau/2} dt + 2 \sum_{n=1}^{\infty} \left(J_{2n}(kva) \int_{-\tau/2}^{+\tau/2} \cos 2n\omega t \, dt \right)$$
$$+ 2j \sum_{n=0}^{\infty} \left(J_{2n+1}(kva) \int_{-\tau/2}^{+\tau/2} \sin(2n + 1)\omega t \, dt \right) \tag{7.6}$$

If the exposure time τ is much greater than the period $2\pi/\omega$ of the oscillations, the two integrals between the square brackets are equal to zero, and the amplitude in the spectrum of the negative is proportional to the Bessel function of order zero $J_0(kva)$. Fringes somewhat similar to the Young's fringes of the experiment of Burch and Tokarski are obtained, but their intensity, which is proportional to $J_0^2(kva)$, decreases faster away from the central fringe. Although we considered the example of Fig. 103, the preceding calculation is more general and may be applied to any configuration. When the diffuse surface A vibrates about an axis contained in its plane, the speckle in a plane E located at a distance d is laterally shifted and the foregoing calculations apply with $f(t) = 2\theta d$. The vibrations of a diffuse object may be detected visually by means of the apparatus shown in Fig. 105, by observing plane H, which is not quite focused on the image of the diffuse object. The areas of the object that are practically stationary will yield a visible speckle. On the other hand, in the vibrating regions the contrast of the speckle is weak, because of the superposition in intensity of multiple speckle patterns. Interesting information may be obtained by means of a photographic recording of the phenomenon followed by a spatial filtering of the negative (Chiang and Juang, 1976a). As previously stated, the greater the defect of focus, the greater the area of the object that is integrated. This inconvenience may be removed by means of the apparatus shown in Fig. 108 (Eliasson and Mottier, 1971). A ground glass G illuminated with a laser projects a speckle pattern on the diffuse surface A whose vibrations are to be studied. A lens O is used to form an image of surface A on the plane of observation E. The phenomenon may be observed either visually or by means of photography. Any displacement of A relative to the speckle produced by G will change the structure of the speckle observed in the plane (see Section 2.8). The stationary parts of the object yield a speckle with high contrast, and the vibrating regions yield speckle with little or no contrast. The sensitivity of the method depends

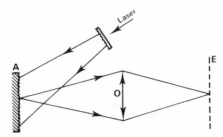

Fig. 108   Study of the vibrations of a diffuse object A illuminated with an auxiliary speckle pattern.

on the dimensions of the speckle projected on A by G, assuming that the lens O has an aperture great enough to resolve this speckle. Experience shows that an appropriate choice of the grain of the ground glass allows the loss of light to remain within reasonable limits.

The vibrations of a diffuse object may also be studied by means of a reference wave (Archbold *et al.*, 1970; Stetson, 1971). The object is illuminated with a laser and the speckle is observed by focusing on the object. A uniform coherent background from the same laser is superimposed on the speckle pattern by means of a beam splitter. Any axial displacement or change of the orientation of the diffuse surface modifies the structure of the resulting speckle, because of interference between the speckle pattern produced by the object and the coherent background.

Note that the apparatus of Fig. 134 described for surface roughness measurements in Chapter IX may also be used to study axial vibrations without being affected by lateral displacements of the diffuse objects (Leger, 1976).

The very simple apparatus of Fig. 109 allows the detection of the vibrations of a diffuse object, when they consist of a change of orientation. The diffuse object that is illuminated with a laser oscillates, for instance, between two positions A_1 and A_2. An image of the diffuse object is formed at A' by means of a lens O. A plane parallel birefringent plate Q cut at 45° from the axis is interposed before or after lens O. Now let the average plane of the object be in a plane such as A_1 in Fig. 110. Because of the birefringent plate Q, the two points M_1 and M_1' of A_1 are superimposed in the image A'. These two points interfere in the image A', where their path difference Δ is caused on the one hand by the difference of the thicknesses of the object at M_1 and M_2, and on the other hand by the path difference Δ_Q produced by the birefringent plate Q. If the diffuse object rotates by a small angle θ, M_1 goes to M_2 and M_1' goes to M_2'. The preceding path difference $\Delta \pm \Delta_Q$ increases or decreases by δ. The variation δ of the path difference produces

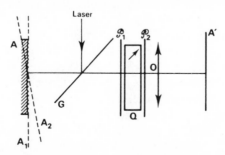

Fig. 109 Study of the vibrations of a diffuse object A by means of a birefringent plate Q.

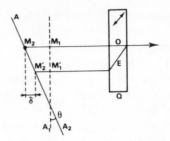

Fig. 110 Variation of the path difference caused by a change in the orientation of the object A.

a change in the structure of the speckle in the image A'. In the parts of the object that do not vibrate, the speckle does not change. Two successive exposures are made with exposure times that are great compared to the period of vibrations, and the photographic plate H is given a small translation between the two exposures. After development, the negative is observed in an apparatus similar to that of Fig. 94. The parts of the object that did not move are eliminated by means of spatial filtering. The areas that were vibrating are visible.

The observation may be made visually in real time. In the stationary areas, the speckle has good contrast. In the moving areas, the contrast of the speckle is weak or even zero, because of the intensity superposition of uncorrelated speckle patterns.

7.10   The Study of the Variations of the Slopes of a Diffuse Object

Consider a polished reflecting surface that is not perfectly planar. In classical interferometry, the slopes of the surface are easily observed by means of the differential method. The wave reflected from the surface is split by an interferometer into two laterally shifted coherent waves. The slopes of the reflecting object may be deduced from the interference between those two waves. It is not possible to apply the same technique to diffuse objects because of the decorrelation of the same speckle pattern between two arbitrary regions. But the method may be applied if we are content to study the variation of the slopes of a diffuse object under deformation.

Consider the diffuse object A shown in Fig. 111. It consists of a surface that oscillates about two average planes $M_1 M_2$ and $M_2 M_3$ with an angle θ between them. We assume that under an arbitrary force the deformation of the object consists only of a change of the angle θ. The lens O forms an

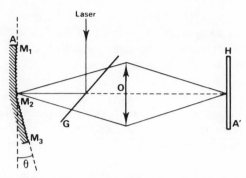

Fig. 111   Study of the changes of the slopes of a diffuse object *A*.

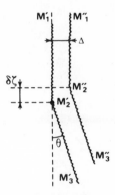

Fig. 112   Variation of the path difference caused by shifting the doubled images by $\delta\zeta$.

image A' of A on the photographic plate H. By means of an interferometer not shown on Fig. 111, two slightly shifted images of the object A are obtained on H. These two images are represented in Fig. 112, where the shift and the path difference produced by the interferometer have been considerably exaggerated. In region $M_1'M_2'$ the path difference between the average planes $M_1'M_2'$ and $M_1''M_2''$ is equal to Δ. The path difference Δ fluctuates because of the irregularities of the surface. In the region $M_2'M_3'$ the path difference between the average planes $M_2'M_3'$ and $M_2''M_3''$ is equal to $\Delta \pm \theta\,\delta\zeta$, where $\delta\zeta$ is the lateral shift (in Fig. 112 the plus sign must be used). In area $M_1'M_2'$ of the image, the observed speckle pattern is a result of interference between the two images $M_1'M_2'$ and $M_1''M_2''$, taking into account the fluctuating path difference Δ. In the area with slope θ, that is, in area $M_2'M_3'$, the observed speckle results from interference between the two images $M_2'M_3'$ and $M_2''M_3''$, taking into account the fluctuating path difference $\Delta + \theta\,\delta\zeta$. A photograph of the image of the diffuse object is recorded. Before the second exposure on the same plate, let the angle θ take another value θ', the variation $\theta - \theta'$ being due to any arbitrary cause. Let the photographic plate be given a small translation ζ_0 between the two exposures. In the region of the image corresponding to $M_1'M_2'$, there are two identical speckle patterns shifted by the distance ζ_0 relative to each other. In the region of the image corresponding to $M_2'M_3'$, because of the variation $(\theta - \theta')\,\delta\zeta$ of the path difference, the resulting speckle has no correlation with the speckle recorded in the same area during the first exposure. After development, the negative is observed as shown in Fig. 94. The slit filter in the focal plane of lens L eliminates the fringes given by the stationary region $M_1'M_2'$, but the light corresponding to region $M_2'M_3'$, which yields no fringes because of

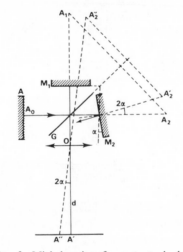

Fig. 113   Use of a Michelson interferometer to double the image.

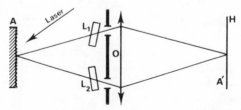

Fig. 114   Use of two parallel glass plates to produce the doubled image.

Fig. 115   Use of four apertures for observation in two perpendicular directions.

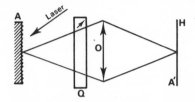

Fig. 116   Use of a birefringent plate Q to produce the doubled image.

the decorrelation of the recorded speckle patterns from the two exposures, goes through. In the image of the negative only the regions where the slopes have changed are visible. The dark regions correspond to the areas of constant slope.

A Michelson interferometer may be used to produce the shift $\delta\zeta$ by changing very slightly the orientation of one of the mirrors (Leendertz and Butters, 1973). In Fig. 113, mirror M_2 is inclined by an angle α, which causes a shift $A'A''$ equal to $2\alpha d$, where d is the distance between the image plane A' and the lens O.

Another type of interferometer has been used by placing two plane parallel glass plates in front of the two apertures of Duffy's configuration shown in Fig. 96 (Hung and Taylor, 1973). By changing slightly the orientation of one of the two plates L_1 and L_2 shown in Fig. 114, two shifted images may be obtained in plane A'. If four apertures (without plates) are used as shown in Fig. 115 instead of two apertures, a lateral shift is produced in two perpendicular directions (Hung and Taylor, 1973) when the recording plate is given a slight defect of focus. This allows the determination of the two perpendicular components of the variations of the slopes.

A simple birefringent plate cut at 45° to the axis may be used as an interferometer. A lens O is used to form an image of the diffuse object A on the plate H as shown in Fig. 116. A birefringent plane parallel plate Q is placed either before or after the lens O. It yields two shifted images in a plane H. The plate Q must be placed between two polarizers not shown on Fig. 116 for the two images at H to be coherent. This simple system is extremely stable.

CHAPTER VIII

Speckle in Astronomy

8.1 The Image of a Star at the Focus of a Telescope in the Presence of Atmospheric Turbulence

In principle, the image of a star at the focus of a telescope is the diffraction pattern (the Airy disk shown in Fig. 3) of the pupil of the telescope. Exceptional atmospheric conditions are required to observe a perfect diffraction pattern: the light received by the telescope from the star must be a plane wave. In general this is not the case in practice, and the wave front may be distorted by atmospheric turbulence. The wave incident on the telescope has irregularities with sizes varying from a few centimeters to some tens of centimeters. The profile Σ shown in Fig. 117 represents the wave front

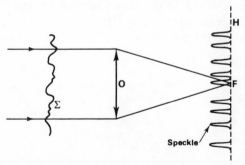

Fig. 117 Speckle at the focus of a telescope in the presence of atmospheric turbulence.

111

incident on the telescope at a given instant. Of course, the shape of the wave front changes very rapidly with time. For this reason the image of a star seen through the telescope by the eye generally has the appearance of a fuzzy spot that changes constantly and whose structure does not look like the Airy diffraction pattern.

Consider the photograph of the image of a star whose apparent diameter is very small compared to the diffraction pattern of the objective O of the telescope (the star is not resolved by the telescope). The recording is made through a monochromatic filter and in the presence of atmospheric turbulence. The sensitivity of the detector is assumed to be sufficiently great to allow very short exposure times, for instance $\frac{1}{100}$ sec. Under these conditions, the wave front Σ may be considered to be "frozen." It practically does not change during such a short exposure time. Because of the irregular shape of the wave front Σ, the image recorded at H consists of a large number of small bright spots distributed at random, the smallest spots having a diameter equal to that of the diffraction pattern of the total aperture of the objective of the telescope. This set of spots is a speckle pattern. This pattern is represented in Fig. 118 as a set of spots all with the same diameter. If a number of photographs are taken, the spots keep the same mean diameter, but their spatial distribution on H changes, and there is no correlation between the recordings. If a number of recordings are made on the same photograph by using a long exposure time, this fine structure disappears. The superposition of all those different pattern yields a much larger fuzzy spot. The advantage of a short exposure time is obvious: because the speckle obtained is formed approximately by diffraction patterns having a diameter equal to that of the diffraction spot of the whole aperture, the information given by the telescope is not lost.

Consider a star with an apparent diameter that is not very small compared to the diameter of the diffraction pattern of the objective O of the telescope, that is, the star is resolved. In the focal plane of the lens O, let the disk A represent the diffraction pattern and the disk S the star, as illustrated in Fig. 119. Each incoherent point of the disk S yields a speckle pattern, which

Fig. 118 Speckle consisting of small spots with dimensions of the order of the size of the diffraction spot of the telescope.

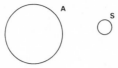

Fig. 119   Simplified representation of the diffraction pattern *A* of the telescope and of the geometrical image *S* of a star.

Fig. 120   After the convolution of *S* with the speckle, the speckle spots spread out.

may be represented at a given instant by Fig. 118, and which we shall represent by a function *D* of two coordinates in the focal plane of the telescope. Finally, the speckle recorded with a short exposure time in the focal plane of the telescope is the superposition of the intensities of all the speckle patterns identical to *D*, which correspond to all the incoherent points of *S*. The spatial distribution of the speckle spots remains the same as that of speckle *D*, for the unresolved star, but each spot becomes larger. The speckle pattern recorded in the focal plane of the telescope is the convolution of image *S* with speckle *D*, which may be written

$$S \otimes D \tag{8.1}$$

The structure of this new speckle pattern is represented in Fig. 120. Figures 118 and 120 are only schematic, and intended to illustrate the phenomena. The use of speckle in astronomy and its applications to the study of double stars and to the measurement of the apparent diameter of stars is due to the French astronomer A. Labeyrie. By its elegance, simplicity, and originality, the work of Labeyrie represents one of the most beautiful contributions of optics to astronomy during the last century.

8.2   The Study of Double Stars at the Focus of a Telescope in the Presence of Atmospheric Turbulence*

Consider a double star and assume for the moment that the two components have equal brightness; because of atmospheric turbulence, each

* See Labeyrie (1974).

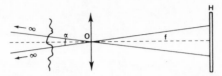

Fig. 121 Two stars of the same magnitude yield two identical shifted speckle patterns at the focus of a telescope.

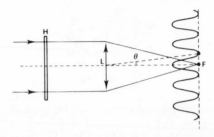

Fig. 122 Spectrum of the negative recorded in the experiment of Fig. 121.

component yields a speckle pattern. Let the phenomenon be recorded as before with a very short exposure time through a monochromatic filter. As illustrated in Fig. 121, two shifted identical speckle patterns are obtained on the photographic plate. If f is the focal length of the objective lens of the telescope, the shift is equal to $f\alpha$, where α is the angular separation of the two components of the double star. This brings us back to the experiment of Section 4.1. After development, the spectrum of the photographic plate H is observed in the focal plane of a lens L, as shown in Fig. 122. As we have seen in Section 4.1, straight parallel fringes are obtained with a maximum contrast equal to unity, and with the angular fringe spacing θ equal to λ/ζ_0, where ζ_0 is the shift $f\alpha$ of the speckle pattern. The simple measurement of $\theta = \lambda/f\alpha$ yields the angular separation α of the double star, if the wavelength λ allowed to go through the filter and the focal length f of the telescope are known. In addition, the spatial orientation of the two components may be deduced from the orientation of the fringes.

When the two components do not have the same brightness, the contrast of the fringes in the spectrum goes down. The ratio $m = B_2/B_1$ of the brightnesses of the components of the double star may be deduced from the contrast γ by means of expression (4.11):

$$\gamma = \frac{2m}{1 + m^2} \tag{8.2}$$

By means of iterated experiments, Labeyrie and his co-workers were able to determine the relative displacements of the components on the orbits about the center of mass of the system. The additional use of spectroscopic orbital data together with the preceding results allowed the determination of the complete orbit of the system and of the masses of the two components.

8.3 The Measurement of the Apparent Diameters of Stars by A. Labeyrie's Methods*

In 1921, Michelson and Pease successfully measured the apparent diameter of Betelgeuse and of a few other bright red stars. The 6-m-long beam put before the 2.5-m Mount Wilson telescope was subject to bending, and considering that the optical paths had to be equalized to an accuracy of 1 μm, it is easy to understand the enormous difficulties that these observers had to contend with. A second interferometer, which used a beam 15-m long, was built in 1930 by Pease, but few results were obtained because of the great difficulty in adjusting such an interferometer. In 1960, Hanbury-Brown and Twiss used a new type of interferometer, the intensity interferometer, which allowed the measurement of the correlation of the signals from two photomultipliers that received light from a star. This correlation is proportional to the square of the modulus of the degree of spatial coherence of the two photomultipliers. The apparent diameter of the star is obtained from the degree of spatial coherence as in Michelson's method. Very good resolution may be obtained if the two photomultipliers are far enough apart, which Michelson and Pease could not achieve. But the degree of spatial coherence is related to the Fourier transform of the energy distribution on the star. The correlation of the signals from the two photomultipliers is therefore proportional to the square of the intensity distribution of the image of the star, and the method is limited to bright stars.

A decisive step forward was taken 50 years after Michelson when Labeyrie had the idea of measuring the apparent diameter of stars by simply studying the spectrum of the speckle patterns produced by the stars. We have seen that the speckle produced by a resolved star may be roughly represented by Fig. 120. For an unresolved star, the speckle consists of much finer spots. Of course the speckle changes very rapidly with time, as does the atmospheric turbulence itself. Let the Fourier transform of the recording H of the speckle pattern of a star be studied in the focal plane of a lens L as shown in Fig. 123. When studied under a microdensitometer, the recording is not smooth,

* See Labeyrie (1976).

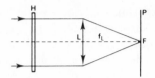

Fig. 123   Observation of the spectrum of a speckle pattern, such as that of Fig. 120.

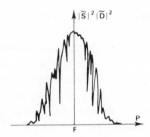

Fig. 124   Discontinuous structure of the spectrum of the speckle.

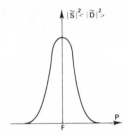

Fig. 125   Superposition of a large number of spectra.

because the speckle consists of small randomly distributed spots. This is illustrated in Fig. 124. Let the star be photographed again. The turbulence and the speckle have changed. Although the diameter of the speckle spots remains the same, only the spatial distribution is changed. The spectrum still has the shape illustrated in Fig. 124, but the discontinuities are not in the same places, because the speckle grains do not occupy the same positions. Now let a large number of spectra of photographs of the speckle patterns of a star be superimposed on the same previously unexposed photographic plate P. The microdensitometer trace then obtained is practically continuous, as illustrated in Fig. 125. For an unresolved star (Fig. 126), the spectrum is wider, because the speckle spots recorded are smaller. Of course we must not superimpose on the same plate the successive recordings of the speckle

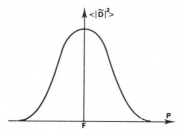

Fig. 126 Spectrum of the speckle of an unresolved star.

patterns and then observe the spectrum of this superposition. This would destroy all the information.

It was shown in Section 8.1 that the speckle recorded on H may be represented by the convolution $S \otimes D$. From Section 4.1 and Eq. (4.1), the amplitude transmitted by H is given by

$$t = a - b(S \otimes D) \tag{8.3}$$

which yields the Fourier transform in plane P shown in Fig. 124:

$$\tilde{t} = a\delta - b(\tilde{S}\tilde{D}) \tag{8.4}$$

The product of the constant a with the delta function represents the image at F of the point source that illuminates the apparatus shown in Fig. 123. In what follows, we consider only the term $\tilde{S}\tilde{D}$.

The curve of Fig. 124 represents the expression

$$|\tilde{S}|^2 \, |\tilde{D}|^2 \tag{8.5}$$

whereas the curve of Fig. 125 represents the expression

$$|\tilde{S}|^2 \langle |\tilde{D}|^2 \rangle \tag{8.6}$$

But the factor $\langle |\tilde{D}|^2 \rangle$ is nothing more than the average spectrum of an unresolved star such as that shown in Fig. 126. Therefore when an unresolved star is under study, a profile similar to that of Fig. 126 will be obtained, which allows the measurement of $\langle |\tilde{D}|^2 \rangle$. This allows the determination of $|\tilde{S}|^2$, which has a profile similar to that shown in Fig. 127. Let the geometrical image S of the star be a uniform disk. In this case, the Fourier transform $\tilde{S}$ of S is well known from diffraction theory: it is an Airy disk with a diameter ε equal to that of the disk S. Figure 127 therefore represents a diffraction pattern whose radius $1.22\lambda f_L/\varepsilon$ may be measured, where f_L is the focal length

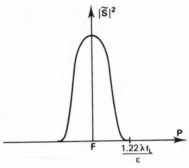

Fig. 127 Comparison of the spectrum of the speckle of a resolved star with the diffraction pattern of a circular aperture.

of lens L of Fig. 123. If f_T is the focal length of the telescope and θ the apparent diameter of the star, the radius is equal to

$$\frac{1.22\lambda}{\theta}\frac{f_L}{f_T} \tag{8.7}$$

The apparent diameter θ of the star may therefore be determined.

To take into account the fact that the star is brighter at the center than at the edge, the disk S must not be considered uniform, and thus the Fourier transform $\tilde{S}$ is not the diffraction pattern of a uniform circular aperture, the Airy disk. It is the Fourier transform of an "apodized" pupil with a diameter S, that is, a pupil with the transparency decreasing towards the edge. For various types of transparencies the diffraction patterns may be easily calculated. By comparing the measured profile of Fig. 127 with the calculated profiles, the profile of the star may be determined.

In 1976, more than 100 stars had been measured with the large Mount Polomar telescope by Labeyrie and his co-workers. Measurements with an accuracy of 0.01″ were made, up to the ninth magnitude.

8.4   The Measurement of the Apparent Diameters of Stars with Multiple Telescopes

The angular resolution of the preceding measurements is limited by the dimensions of the largest existing telescopes. Because it will not be feasible to increase considerably their dimensions, A. Labeyrie thought that it would be much simpler to use two telescopes T_1 and T_2 separated by a distance that would determine the resolution.

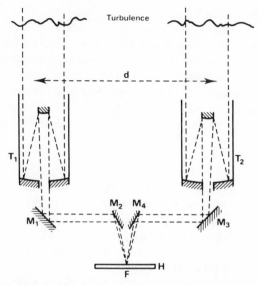

Fig. 128   Recording of the speckle at *H* produced by two telescopes T_1 and T_2.

The apparatus could be that shown in Fig. 128. After going through telescope T_1, the light is reflected by two mirrors M_1 and M_2 and is projected on the detector *H* located at the focus *F*. The light from telescope T_2 follows a symmetrical path. The detector *H* is at the focus of telescope T_2. This experiment is practically that of Section 3.5. At any given instant, the turbulence before telescope T_1 is different from the turbulence before telescope T_2. At the focus *F* of telescope T_1 (reached through the path $T_1 M_1 M_2 F$) the speckle pattern is different from that produced by telescope T_2, through path $T_2 M_3 M_4 F$. The optical paths are equalized in such a way that the two speckle patterns are coherent. Each resulting speckle grain is modulated by straight parallel fringes, as in Fig. 50. The angular distance between two consecutive fringes is equal to $\lambda f / d$, where f is the focal length of the telescope and d is the distance between the axes of the telescopes. The same operations as before are carried out, but it is the fringes that are of concern.

The recording at *H* of one image of this modulated speckle with a sufficiently short exposure time yields $S \otimes D$, where *D* represents the set of fringes of any arbitrary grain of the speckle. This is a one-dimensional convolution in a direction perpendicular to the fringes. All the other operations are the same as before. But because *D* represents fringes that may be very fine if d is very great, considerable resolution may be obtained. The state of present technology should allow resolutions of the order of 10^{-4} seconds of arc. One might fear that large base lines d would require impossible feats of optical path equalization. But the available image detectors are so sensitive that it is possible to use monochromatic filters with bandwidths small enough that the path lengths must only be equalized to about 1 mm.

The first experiment was successfully done at the Meudon observatory with two telescopes having diameters of 25 cm, separated by a distance of 12 m.

When two telescopes are used, the convolution is only in one direction and of course the resolution obtained is in this direction. There is nothing to prevent the experiment from being carried out with a number of nonaligned telescopes. With three telescopes at the vertices of an equilateral triangle, the speckle grains are modulated by a very fine hexagonal structure, if the sides of the triangle are long. The preceding operations will be repeated with this configuration. This will allow the maximum resolution in three directions oriented at angles of 60° to each other.

It may be shown (Liu and Lohmann, 1973) that the image of a star may be reconstructed by combining the recording of the image obtained with a long exposure and the autocorrelation of the image obtained with a short exposure. This method only applies to small objects on a dark background, which is the case for stars.

CHAPTER IX

The Study of Surface Roughness

9.1 Surface Deviations

The defects of a surface compared to the ideal geometry may be classified in the following manner:

(a) Shape deformation. This is a defect of the geometry of the surface as a whole, and is not a defect of the state of the surface. For a surface whose ideal geometry would be a plane, the shape deformation is represented by curve (1) of Fig. 129.

(b) Surface roughness. This is mainly due to the gouging of the surface by the tools preparing the surface. Surface roughness generally consists of random defects, but the surface may also have periodic defects. The defects have a high spatial frequency. In this chapter, only surface roughness due to random defects such as those shown in curve (2) of Fig. 129 will be considered.

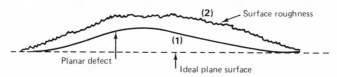

Fig. 129   Deviations of a real surface from an ideal plane surface.

A certain number of criteria have been defined to describe the state of a surface. Among the statistical criteria is the mean deviation, which is a measure of the deviation of the surface relative to an average profile. Another statistical measurement is the standard deviation σ. The generally accepted relationship between the mean deviation R_m and the standard deviation σ is $\sigma = 1.11 R_m$. In fact, the proportionality factor depends on the type of surface under consideration. The value 1.11 is an average value for the set of surfaces that are dealt with in mechanics. Depending on the procedure used, the proportionality factor may vary between 1.0 and 1.3. In most cases a proportionality factor equal to unity is accurate enough.

9.2   The Use of Speckle to Study Surface Roughness

Surface profiles are generally studied by means of a stylus that is in contact with the surface. The vertical displacements of the sensor are transformed into voltage variations and then amplified and recorded. The low frequencies are filtered out, and the remaining high frequencies correspond to the surface roughness. The main disadvantage of this kind of apparatus is the contact between the point of the stylus and the surface, which may deteriorate the surface. On the other hand, optical techniques do not touch the surface, but only use the light reflected from the surface. The classical optical methods will not be described here; we shall consider only the use of speckle.

Generally the information on the surface roughness is obtained from the speckle by studying the correlation between two speckle patterns obtained from the surface under consideration, either by changing the orientation of the laser beam or by changing the wavelength of the laser beam. Other techniques that relate the speckle contrast to the surface roughness and to the spatial or temporal coherence of the source have been proposed.

The methods that use speckle are well adapted for the measurement of very rough surfaces ($\sigma > 1$ μm), and in certain cases, the direct measurement of a statistical parameter may be carried out in real time.

9.3   Surface Roughness Measurement by the Correlation of Two Speckle Patterns Obtained by Changing the Incidence of a Laser Beam*

The surface under study is illuminated with a plane wave from a laser as shown in Fig. 130, and the two speckle patterns obtained with two different

* See Leger *et al.* (1975).

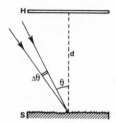

Fig. 130   Recording at *H* of two speckle patterns corresponding to two orientations of the incident beam.

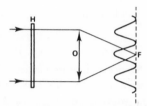

Fig. 131   Display of straight parallel fringes in the spectrum of the negative.

angles of incidence are recorded successively on a photographic plate *H*. The two speckle patterns on plate *H* are therefore translated relative to each other. This is the experiment described in Section 2.4 for a transparent diffuse object, done in this case with a diffuse reflecting object. We have seen that changing the angle of incidence does not produce any decorrelation under conditions that have been described previously. The situation here is the same. If the change of the angle of incidence is small enough, the speckle will simply be translated by a distance ζ_0 given by

$$\zeta_0 = d \cos \theta \, \Delta\theta \qquad (9.1)$$

where *d* is the distance between the scattering surface *S* and the photographic plate, and θ is the angle of incidence of the laser beam on surface *S*. When the change of incidence $\Delta\theta$ increases too much for a given angle θ, the speckle is not only shifted, but its structure changes due to the roughness of the surface *S*. This is the principle of the technique: two successive exposures are made on the same plate *H*, with a change $\Delta\theta$ of the angle of incidence of the beam that illuminates *S*. After development, the negative *H* is observed with parallel light by means of the classical apparatus that was used many times, for example, with the apparatus of Fig. 55, which is reproduced in Fig. 131. In the focal plane of the lens *O* are obtained straight parallel fringes with an angular spacing λ/ζ_0 and with a contrast that depends on the roughness of the surface.

It may be shown that for very rough metal surfaces ($\sigma > 1 \ \mu$m) the contrast γ of the fringes defined by Eq. (4.11) is given by the expression

$$\gamma = \exp\left[-\left(\frac{2\pi\sigma}{\lambda}\sin\theta\,\Delta\theta\right)^2\right] \tag{9.2}$$

where σ is the standard deviation of the surface roughness. The value of σ may therefore be obtained from the measurement of the contrast γ of the fringes. Experience has shown that a relative precision $\Delta\sigma/\sigma$ of 8% may be attained. It is possible to measure large values of σ, but the measurement of irregularities greater than 30 μm has no practical interest. It is interesting to note that this method may yield accurate results even when it is applied to dielectric rough surfaces.

9.4 The Real-Time Measurement of Surface Roughness by the Amplitude Correlation of Two Speckle Patterns Corresponding to Two Orientations of the Laser Beam*

In the preceding sections, the surface roughness was obtained from the measurement of the correlation of the intensities of two speckle patterns. Real-time measurements may be made if the intensity correlation is replaced by an amplitude correlation. The surface under study is illuminated by two coherent plane waves from the same laser. The incident beam is divided by means of an interferometer represented in Fig. 132 by a beam splitter G and a mirror M. For an angle of incidence θ, there is a speckle pattern D that may be observed in direction θ'. For an angle of incidence $\theta + \Delta\theta$, there is another speckle pattern D', which is related to D by a rotation $\Delta\theta'$

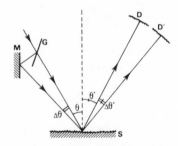

Fig. 132 Recording of two coherent speckle patterns corresponding to two orientations of the incident beam.

*See Leger (1976).

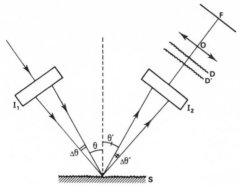

Fig. 133   Use of two Michelson interferometers to observe the roughness of surface *S* in real time.

and which has with the latter a certain degree of correlation. The angle of rotation is

$$\Delta\theta' = \frac{\cos\theta}{\cos\theta'}\,\Delta\theta \tag{9.3}$$

In order to measure the correlation between two speckle patterns *D* and *D'*, which is related to the surface roughness, the two speckle patterns *D* and *D'* are superimposed by means of a two-beam interferometer. The apparatus is shown in Fig. 133, where two identical interferometers I_1 and I_2 may be used. Two Michelson interferometers may be used, the second interferometer not being adjusted for zero path difference, so that the two speckle patterns *D* and *D'* are shifted along the axis as shown in Fig. 133. At the focus of lens *O* are observed rings whose contrast is related to the correlation of the two speckle patterns *D* and *D'*, that is, to the surface roughness. For a very rough metal sample, the contrast of the fringes may be shown to be given by the expression

$$\gamma = \frac{1}{2}\exp\left[-\left(\frac{2\sqrt{2}\pi\sigma}{\lambda}\sin\Delta\theta\right)^2\right] \tag{9.4}$$

which may be compared to expression (9.2). Although it is not necessary, there is advantage in taking $\theta = \theta'$, which yields more light and makes the adjustments easier. If the interferometer is adjusted for a zero path difference, there is a uniform field at *F*. If a photoelectric detector is put in this plane and the phase difference between the two interfering beams is

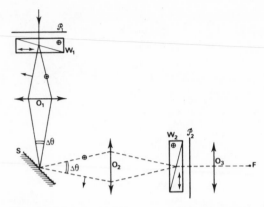

Fig. 134 Use of two Wallaston prisms for the real-time observation of the roughness of surface *S*.

modulated by some means, the intensity at *F* is modulated. The amplitude of the signal modulation given by the detector yields the same result as the fringe contrast.

In the case when the angles $\Delta\theta$ and $\Delta\theta'$ are fixed once and for all, the two Michelson interferometers may be replaced by two Wallaston prisms, which yields a stable and practical apparatus. This is illustrated in Fig. 134. The lens O_1 projects the fringes from the Wallaston W_1 on the surface *S* under study. The lens O_2 projects these fringes on the Wallaston W_2 in such a way that there is compensation, the two Wallaston prisms being between two polarizers $\mathscr{P}_1$ and $\mathscr{P}_2$. This means that the two orthogonally polarized rays split by Wallaston W_1 are recombined after going through Wallaston W_2. Let a small axial translation be given to Wallaston W_2. In the focal plane *F* of lens O_3, which is placed after Wallaston W_2, straight parallel fringes are observed with a contrast related to the surface roughness of *S* as before. Instead of displacing Wallaston W_2 to make the fringes appear, the compensation may be conserved. The field is uniform in the focal plane *F*, and when the phase difference between the two rays that are polarized at right angles is modulated, the intensity at *F* is modulated. A photoelectric detector located at *F* allows the measurement of the amplitude modulation, which gives a result identical to that given by the contrast of the fringes in the preceding case.

Contrary to expectations, the depolarization produced by metal surfaces with roughness within practical and interesting bounds is weak and does not interfere with the measurements. Of course, the preceding apparatus may not be used for surfaces that completely depolarize the light.

9.5   Surface Roughness Measurement by the Correlation of Two Speckle Patterns Obtained with Two Wavelengths*

Instead of changing the orientation of the incident beam as previously described, the wavelength may be changed instead. Of course, this does not allow real-time measurements. The apparatus is shown in Fig. 135. The surface S under study is illuminated with laser light having wavelength λ. An image of S is projected on the photographic plate H by means of a lens O. A speckle pattern corresponding to wavelength λ is therefore recorded. Before the second exposure, the photographic plate H is given a small translation ζ_0. The surface S is then illuminated with light having wavelength $\lambda + \Delta\lambda$ and the speckle corresponding to this new wavelength is recorded at H. Because of the roughness of surface S, the two speckle patterns are somewhat decorrelated. The decorrelation may be shown to be complete if

$$\sigma \geq \frac{\lambda^2}{2\,\Delta\lambda} \tag{9.5}$$

where σ is the standard deviation.

After development, the negative H is illuminated with parallel light and the fringes are observed at the focus of a lens, with the same procedure as before. The fringes disappear when the relation (9.5) is satisfied. The value of $\Delta\lambda$ is increased until the fringes disappear, and this allows the measurement of σ.

The intermediary optics, that is, the lens O of Fig. 135, may be suppressed by recording the speckle at a finite distance as shown in Fig. 136 (Mendez and Roblin, 1975a). Two successive exposures are made on H. During the first exposure the surface S is illuminated with light of wavelength λ. Before the second exposure, the photographic plate H is given two translations: (a) a lateral translation ε_0, which later allows the observation of the fringes in the spectrum of the negative; and (b) an axial translation ε to compensate for the variation $\Delta\lambda$ as seen in the experiment of Section 2.5 and illustrated

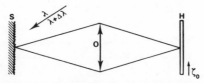

Fig. 135   Surface roughness measurements with two wavelengths.

* See Tribillon (1974).

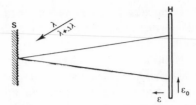

Fig. 136 Study of the roughness of S by recording a speckle pattern at a finite distance with two wavelengths.

in Fig. 36. Under these conditions two identical speckle patterns are obtained on H if condition (2.10) is satisfied, that is, if

$$\sigma \ll \frac{\lambda^2}{\Delta\lambda} \tag{9.6}$$

If this condition is not satisfied, the two speckle patterns are more or less decorrelated. After development, the spectrum of the negative is observed with the usual apparatus, and the correlation of the speckle patterns, that is, σ, may be deduced from the contrast of the fringes.

9.6   Surface Roughness Measurements with a Source Having a Wide Bandwidth*

When a rough surface is illuminated with spatially coherent light, the speckle has the maximum contrast. If the temporal coherence is reduced while the spatial coherence is maintained, the contrast goes down. For a given coherence length, the influence of the surface roughness on the contrast

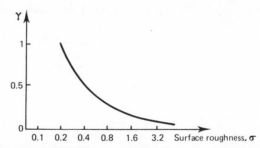

Fig. 137 Relationship between the roughness and the speckle contrast when a source of white light is used (Sprague, 1972).

* See Sprague (1972).

of the speckle is particularly great when the variations of the thickness of the surface are of the order of magnitude of the coherence length. This principle may be used for the evaluation of surface roughness. Experience has shown that the relation between the speckle contrast and the surface roughness is almost independent of the technique used to prepare the rough surface. For white light with a coherence length approximately equal to 1.5 μm, the surface roughness may be evaluated by measuring the contrast of the speckle if the roughness is between 0.2 and 0.3 μm, as shown in Fig. 137. For instance, the contrast of the speckle may be determined by projecting an image of the surface under consideration on a small aperture behind which is a detector. When the surface is translated in its plane, the profile of the speckle is measured. The diameter of the aperture must be smaller than the diameter of the smallest speckle grains.

9.7   Surface Roughness Measurements with Partially Coherent Light*

When a transparent diffuse object is illuminated with a laser, that is, with a spatially and temporally coherent source, the contrast of the speckle in the image of the object is a maximum. If the diffuse object is illuminated by a source having temporal coherence, but with partial spatial coherence, the contrast of the speckle is reduced. In this case the contrast of the speckle is affected by the roughness of the diffuse surface, and the roughness may be measured from the contrast of the speckle. The experiment is described in Fig. 138. A condenser C projects an image of the monochromatic source S on a small aperture T located at the focus of a lens L. The light from this illuminates the diffuse transparent object G, for instance, a ground glass. The two lenses O_1 and O_2 form an image of G on a photoelectric detector R. In front of the detector R there is a screen with an aperture smaller than the diameter of the speckle grains. The diaphragm P, which is located in the focal plane of the lens O_1, allows the size of the speckle grains on the detector

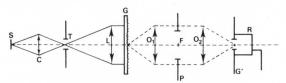

Fig. 138   Measurement of the roughness of G by means of spatially coherent light. $\lambda = 0.5$ μm, $L_c = 1.5$ μm.

* See Fujii and Asakura (1975).

to be changed. When the detector is moved in the image plane G', the intensity profile of the speckle may be determined.

The degree of spatial coherence in plane G may be determined from the Zernike–Van Cittert theorem. According to this theorem, the degree of coherence of two points located in plane G is expressed as the Fourier transform of the intensity distribution in the plane of aperture T. This is equal to the Fourier transform of the aperture T if it is illuminated with coherent light. The amplitude of this Fourier transform is equal to

$$U = \frac{2J_1[K(a/f)\rho]}{K(a/f)\rho} \qquad \left(K = \frac{2\pi}{\lambda}\right) \tag{9.7}$$

where J_1 is the Bessel function of order one, a is the radius of aperture T, f, the focal length of lens L, is the distance TL, and ρ is the distance between the two points under consideration. Let the degree of spatial coherence at G be defined as the value of ρ that corresponds to the first zero minimum of U, that is,

$$\rho = \frac{1.22\lambda f}{2a} \tag{9.8}$$

For a given value of ρ, the variation of the speckle contrast as a function of the standard deviation σ, which is a measure of the roughness, is practically linear up to a certain value of σ that depends on ρ. Figure 139 shows the results. In this figure, the contrast γ of the speckle is defined by

$$\gamma = \frac{\sqrt{\langle \Delta I^2 \rangle}}{\langle I \rangle} \tag{9.9}$$

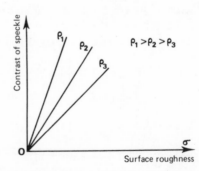

Fig. 139 Contrast of the speckle as a function of the surface roughness (Fujii and Asakura, 1975).

where ΔI represents the intensity variations measured by the detector R of Fig. 138.

For ground glasses with different grains, for $\rho_1 = 444\ \mu$m, the linearity goes from $\sigma = 0$ to $\sigma = 0.3\ \mu$m approximately. For $\rho_2 = 14\ \mu$m, $\rho_3 = 3\ \mu$m, and $\rho_4 = 3\ \mu$m, it goes from $\sigma = 0$ to $\sigma = 0.5\ \mu$m, $\sigma = 1.5\ \mu$m, and $\sigma = 3\ \mu$m, respectively.

CHAPTER X

Various Applications of Speckle

10.1 Differential Interferometry of Transparent Objects by Double-Exposure Speckle Photography

The principle of this experiment (Kopf, 1972b; Mallick and Roblin, 1972) is shown in Fig. 140. A ground glass G, illuminated with a laser, is put in front of the transparent object A under study. The lens O forms an image A' of A on photographic plate H. An image of the speckle produced by G in the object plane A is observed in the image plane A'. Two exposures are made, the object being given a small axial translation between the two exposures. For an area of A where the slope is zero, this shift does not produce any change in the structure or in the position of the speckle at H. However, as shown in Fig. 141, in an area of slope α, the speckle at H is shifted laterally. For example, in Fig. 141, the initial focusing on H is made on

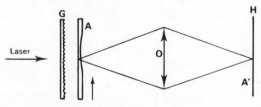

Fig. 140 Study of the slopes of a transparent object A with the speckle pattern produced by G.

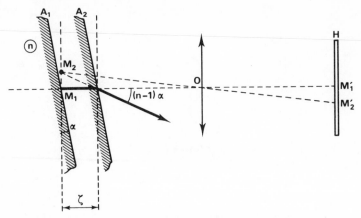

Fig. 141   Change of the path difference under a small axial translation of the object.

a speckle pattern located on a plane through M_1. After the transparent object has been shifted by ζ, point M_1 goes to M_2 and a lateral shift M_1M_2 of the speckle appears. There is a corresponding displacement $M_1'M_2'$ in the image plane. Let n be the index of refraction and α be the slope of the transparent object. The lateral shift M_1M_2 is equal to $(n-1)\alpha\zeta$; if the photographic plate H is given a small lateral translation between the two exposures, the areas of nonzero slope α may be observed by putting the negative on an apparatus similar to that of Fig. 94. The slit filter allows the elimination of the areas with zero slope.

10.2   The Use of Two Wavelengths to Study the Shape of Diffuse Objects*

The methods described in Chapter VII do not allow the observation of the shape of a surface, but only the changes of the shape. Speckle recording affords a solution to the problem by combining the use of an interferometer with the use of two laser wavelengths. Consider the interferometer represented in Fig. 142. For simplicity, the diffuse object A is assumed to consist of two planes A_1 and A_2 separated by a distance ε. The object A is successively illuminated with laser wavelengths λ_1 and λ_2. The image of A given by lens O is observed at plane H. Let the distance ε be small enough so that the images A_1' and A_2' of A_1 and A_2 are in focus. Let Δ_1 be the path difference

* See Ennos (1975).

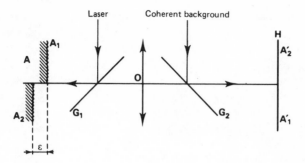

Fig. 142   Study of the average shape of a diffuse surface by using two wavelengths.

at a point of the image A_1' located between the beam scattered by the conjugate point of the object and the coherent beam from the same laser. Let the path difference Δ_1 be m times the wavelength λ_1, that is, $\Delta_1 = m\lambda_1$. If this path difference Δ_1 contains the wavelength λ_2 n times ($\Delta_1 = n\lambda_2$), the coherent background is in phase with the speckle produced at H by A_1 for both wavelengths. The two resulting speckle patterns at H are then practically identical if the difference of wavelengths is small, as discussed in Section 1.9. The path difference satisfies the relations

$$\Delta_1 = m\lambda_1 = n\lambda_2, \qquad \frac{\Delta_1}{\lambda_1} = m, \qquad \frac{\Delta_1}{\lambda_2} = n \qquad (10.1)$$

and if we set $m - n = p$, then

$$\frac{\Delta_1}{\lambda_1} = \frac{\Delta_1}{\lambda_2} + p \qquad (10.2)$$

For another position corresponding, for example, to A_2 the situation is similar if

$$\frac{\Delta_1}{\lambda_1} = \frac{\Delta_2}{\lambda_2} + p + 1 \qquad (10.3)$$

For this new position, the two speckle patterns observed at H with wavelengths λ_1 and λ_2 are again practically identical.

Because $\Delta_2 - \Delta_1 = 2\varepsilon$, from Eqs. (10.2) and (10.3),

$$2\varepsilon = \frac{\lambda_1\lambda_2}{\lambda_2 - \lambda_1} \qquad (10.4)$$

Thus each time that the deviation of the depth of the object A with respect to position A_1 is equal to an integral multiple of the value of ε given in Eq. (10.4), two practically identical speckle patterns are obtained at H. Two successive exposures are made on H, the first with λ_1 and the second with λ_2, and the plate H is given a small translation between the two exposures. By filtering the negative as previously described, the areas of the object whose deviations of depth relative to A_1 satisfy Eq. (10.4) are eliminated from the negative. The contours of these areas appear on the surface of the object A as dark fringes. When $\lambda_1 = 0.488$ μm and $\lambda_2 = 0.486$ μm, the passage from one dark fringe to the next corresponds to a thickness variation in the object equal to 14.2 μm.

10.3   The Use of Speckle to Determine the Transfer Function of an Optical System*

The principle of the experiment is illustrated in Fig. 143. An extended monochromatic source illuminates a random transparency A. This transparency is the photographic recording of a speckle pattern. The lens O under study forms an image of the object A at A'. Because the source S is an extended source, A is illuminated with spatially incoherent light.

The intensity $I(\eta, \zeta)$ at an arbitrary point (η, ζ) of the image A' results from the convolution of the geometrical image $D(\eta, \zeta)$ of A and the intensity $E(\eta, \zeta)$ of the diffraction pattern of lens O:

$$I(\eta, \zeta) = D(\eta, \zeta) \otimes E(\eta, \zeta) \tag{10.5}$$

After development, the amplitude transmittance of the negative is proportional to the recorded intensity $I(\eta, \zeta)$. When the negative is observed on an apparatus such as that shown in Fig. 94, the amplitude in the focal plane of a lens L placed after the negative H is proportional to the Fourier transform of $I(\eta, \zeta)$

$$\tilde{I}(u, v) = \tilde{D}(u, v)\tilde{E}(u, v) \tag{10.6}$$

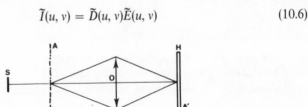

Fig. 143   Study of the transfer function of a lens O.

* See Courjon *et al.* (1975).

The Fourier transform $\tilde{E}(u, v)$ of the diffraction pattern of the lens O is the transfer function of this lens. What is observed in the focal plane of lens L is therefore the square of the modulus of the transfer function on lens O. Of course, the transfer function of lens O is obtained for the conditions under which this lens is used. In the case of Fig. 143, the lens O is working for the antiprincipal points if the distances AO and OH are equal.

The phase of the transfer function may be obtained by means of the double-exposure technique. Two successive exposures are made on photographic plate H; in the first exposure the recording is made as shown in Fig. 143. In the second exposure the lens O is replaced by a reference lens O' that is assumed perfect, and the photographic plate H is given a small translation in its plane between the two exposures. The spectrum of the negative is then observed in the usual manner. If the lens O has no aberrations, straight parallel fringes are observed in the spectrum. The two lens O and O' being identical, two identical speckles shifted with respect to each other, because of the translation of H between the exposures, are recorded on H. But if the lens O has aberrations, the fringes of the spectrum are not straight and parallel, and the phase of the transfer function of lens O may be determined from the deformation of those fringes.

The delicate point of this method is that the lenses O and O' must be accurately positioned on the apparatus of Fig. 143.

10.4   The Study of the Aberrations of an Optical System
by Speckle Photography*

This experiment is illustrated in Fig. 144. An amplitude diffuser A (the photography of a speckle pattern) is illuminated at normal incidence with a collimated beam from a laser and its image is formed on H by means of the lens O whose aberrations are to be studied. For simplicity, we shall consider

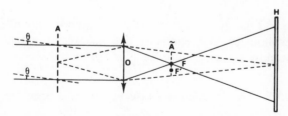

Fig. 144   Study of the spherical aberration of a lens O by recording the speckle with two orientations of the incident beam.

*See Roblin *et al.* (1977) and Tanner (1969).

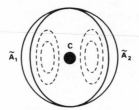

Fig. 145 Interference between two reconstructed images yields the derivative of the function that represents the aberration (the differential method).

the case of a lens with only spherical aberration. The recorded image on photographic plate H may be considered to be a Fourier hologram. The reference source is the direct image F of the light source and the object is the spectrum $\tilde{A}$ of the diffuser A. The phase variations of $\tilde{A}$ represent the spherical aberration of the lens O. After development, the negative H is illuminated with a parallel beam of light, and H then reconstructs two images of $\tilde{A}$ that are symmetrical with respect to the center. In this case, the two images are identical (the aberration has circular symmetry) and are superimposed, because F is at the center of $\tilde{A}$.

   If the diffuser A is illuminated with a beam at an angle of incidence θ, the hologram at H is recorded with a reference source F' that is not centered on $\tilde{A}$. Upon reconstruction, two images $\tilde{A}_1$ and $\tilde{A}_2$, separated and symmetrical with respect to the center, will be obtained as illustrated in Fig. 145. If the distance between those two images is small, their interference will produce fringes that represent the derivative of the function representing the spherical aberration of the lens O under the conditions of the experiment. This is a differential method.

   In order to operate under appropriate conditions, only a small portion of the negative H is used. In this way, everything happens as if the reference wave convergent upon F' had gone through only a small region of the lens O, which allows the reference wave to be considered as practically free of aberrations.

   Experience and theory have shown that this process may be extended to aberrations without circular symmetry and that the amplitude diffuser A may be replaced by a phase diffuser such as a ground glass.

10.5   The Focusing of a Lens by Means of Speckle*

   It is required to put a film A in the focal plane F of a lens O_1, as illustrated in Fig. 146. The lens is illuminated with parallel laser light that converges

*See Sawatari and Elek (1973).

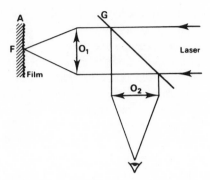

Fig. 146   Focusing of the lens O_1 on the film A by the observation of the speckle produced by A.

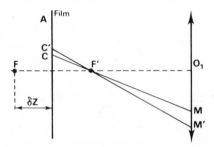

Fig. 147   Relative displacements of the film and of the speckle produced by the lens O_1.

in the film A. The light scattered back from F goes through lens O_1 and is reflected on the beam splitter G to the eye located at the focus of an auxiliary lens O_2, to observe the lens O_1. In Fig. 147 the film is shown at a small distance δz from the focus F of lens O_1. A small area Φ of the film is illuminated by the laser beam and diffracts the light. It produces a speckle pattern in the plane of lens O, which is assumed to be a thin lens. For example, the point M is observed.

Consider the ray CM through point F' symmetrical to F with respect to the plane of the film. If the film moves upwards by a distance ζ, point C goes to C' and point M to M'. The speckle that was previously at M is now at M'. If δz is small, then

$$\overline{MM'} = \frac{f\zeta}{\delta z} \tag{10.7}$$

where f is the focal length of the lens O. If the film moves at speed v, the speckle at M moves in the opposite direction with a speed $vf/\delta z$. When the focus F is between the film and the lens, the displacement of the speckle is in the same direction as that of the film. On the other hand, the speckle grains at M have a diameter approximately equal to λ/Φ, where $\Phi = D\,\delta z/f$ is the diameter of the small area illuminated on A, and D is the diameter of the lens O_1. When δz decreases, that is, as the film approaches the focal plane, diameter of the speckle grains increases, and in principle goes to infinity together with the speed of the speckle displacement. This allows an accurate determination of the position of focus.

In fact, the focus F of the lens O_1 is not a geometrical point, but a small diffraction pattern that depends on the wavelength and the aberrations of lens O_1, which makes it possible to study those aberrations too.

10.6   Laser Speckle for Determining the Ametropia
of the Eye*

A screen is illuminated with a laser and the eye looking at the screen sees a speckle pattern. Let the eye be normal and focused on infinity as shown on Fig. 148. A ray as IJ reaches the focus F_1 of the eye O. If the eye moves upwards to OF, the ray IJ has the position shown in Fig. 149, and after being refracted again falls on F. Let us reconsider Fig. 148 with a myopic eye illustrated in Fig. 150. The ray IJ reaches point A on the retina. If the eye moves upwards, the ray IJ has the position shown on Fig. 151, and the speckle pattern also moves upwards. Figure 152 illustrates the case for a

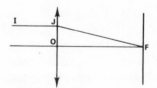

Fig. 148   Speckle produced at F on the retina of a normal eye.

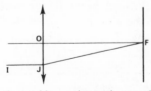

Fig. 149   For a normal eye, the speckle remains stationary when the head is moved vertically.

* See Mohon and Rodeman (1973).

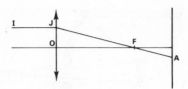

Fig. 150 Case of a myopic eye.

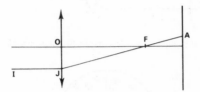

Fig. 151 For a myopic eye, the speckle moves upwards when the head is moved up.

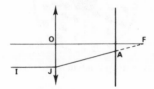

Fig. 152 For a hypermetropic eye, the speckle moves downwards when the head is moved up.

hypertropic eye. If the eye moves upwards, the speckle on the retina moves downwards. It is therefore possible to correct the eye by choosing glasses such that the speckle remains stationary when the head is moved.

10.7 The Use of the Double Exposure of a Random Distribution of Intensity to Measure Atmospheric Turbulence*

Let the eye be at the focus of a telescope pointing at a star as shown in Fig. 153. The eye looks at the entrance pupil of the telescope and not at the star. Because of atmospheric turbulence, the entrance pupil is not uniform. A random distribution of luminous spots that move rapidly are seen as illustrated in Fig. 154. The propagation speed of the spots are a direct reflection of the speed of the wind in the turbulent zone responsible for the twinkling of the star. Despite the fact that the spots are not very small, they may

---

* See Martin *et al.* (1975).

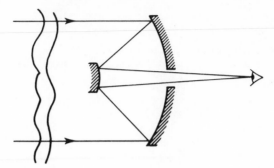

Fig. 153   Observation of the pupil of a telescope in the presence of atmospheric turbulence.

Fig. 154   "Flying shadows" are a random distribution of small bright spots.

be considered as speckle and the double-exposure technique may be used. The phenomenon is recorded with a high-speed cinematographic film, because of the speed of the luminous spots. The film is focused on the entrance pupil of the telescope by means of the camera lens. By means of a special shutter, two very short exposures separated by a time interval of a few milliseconds are superimposed on the same film. If the luminous spots move without deformation during the time interval t, there are on each frame two identical random distributions shifted by a distance that depends on t. After development, the spectrum of the negative is examined on an apparatus similar to that of Fig. 94. Straight parallel fringes are observed, and the fringe spacing and the orientation of the fringes yield the speed and the direction of the wind in the turbulent zone.

10.8   The Measurement of Motion Trajectories by Speckle Photography*

Let a diffuse object be subjected to a lateral displacement, and for simplicity consider only one-dimensional phenomena. The displacement of the object is parallel to the ζ axis.

* See Lohmann and Weilgelt (1975a, 1976).

The intensity $D(\eta, \zeta)$ of the speckle pattern in the image of the object is photographically recorded. When two successive exposures are made, if the object has moved a distance $2\zeta_1$ in the image plane, the convolution given by expression (4.3) is recorded. In this case the trajectory of the object is represented by

$$p(\zeta) = \delta(\zeta + \zeta_1) + \delta(\zeta - \zeta_1) \qquad (10.8)$$

Let a third exposure be made on the same plate, the photographic plate being given a translation ζ_0 much greater than $2\zeta_1$. The recorded intensity is

$$D(\eta, \zeta) \otimes [p(\zeta) + \delta(\zeta - \zeta_0)] \qquad (10.9)$$

After development, the spectrum of the negative is observed; it has an amplitude proportional to

$$\tilde{D}(u, v)[\tilde{p}(v) + \exp(-jKv\zeta_0)] \qquad (10.10)$$

and an intensity

$$|\tilde{D}(u, v)|^2[\tilde{p}(v)\exp(jKv\zeta_0) + \tilde{p}^*(v)\exp(-jKv\zeta_0) + 1 + |\tilde{p}(v)|^2] \quad (10.11)$$

This expression is recognized as the Fourier hologram of an object $p(\zeta)$. In the foregoing example, the object $p(\zeta)$ simply consisted of two points. After recording this spectrum on a photographic plate H, the spectrum of H is observed: just as in Fourier holography, it reconstructs two groups of two points symmetrical with respect to the center of the spectrum, as illustrated in Fig. 155. Two "images" of the object $p(\zeta)$ are reconstructed on either side of the center of the spectrum. By introducing cosine functions into Eq. (10.11), the Fourier transforms leading to the result of Fig. 155 are easily extracted. This method is general, and it allows the reconstruction of any arbitrary object $p(\zeta)$. The added exposure with $\delta(\zeta - \zeta_0)$ serves as a reference source.

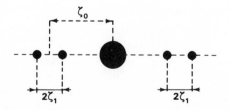

Fig. 155 Reconstruction of the positions of an object by the recording of three speckle patterns.

The procedure may also be applied to an axial displacement if the object is observed at an angle, so that an axial displacement also produces a lateral displacement. It is also feasible to study axial displacements by correlating the speckle patterns corresponding to different positions of the object to speckle patterns corresponding to known displacements of the same object.

10.9   The Determination of the Velocities of Different Parts of a Diffuse Object by Speckle Photography*

The spatial distribution of the velocities of a diffuse object may be studied by illuminating the object with a pulsed laser. An electro-optical modulator, a rotating disk with apertures, or a pulsed laser may be used. Figure 156 represents the type of illumination required: at least two pulses of duration τ separated by a time interval T. It is assumed that the exposure time is small enough that each object point moves in a straight line with a constant speed during the exposure. An image of the diffuse object A is formed on a photographic plate H by means of a lens O as shown in Fig. 157, and for simplicity,

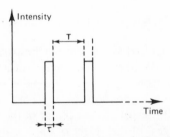

Fig. 156   Modulation of a laser beam required to study the velocity of a diffuse object.

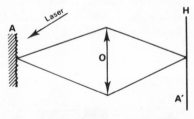

Fig. 157   Recording of the image of an object illuminated under the conditions shown in Fig. 156.

* See Celaya *et al.* (1976).

the displacements of the different parts of the object are assumed to be in a direction perpendicular to the axis of the system. Let $D(\eta, \zeta)$ be the intensity of the speckle in image A'. During the exposure, let an arbitrary part of the object move parallel to the axis ζ. Let $\zeta_0 = V\tau$, where V is the speed of this area. The photographic recording during the time interval τ may be written

$$D(\eta, \zeta) \otimes \text{Rect}\left(\frac{\zeta}{\zeta_0}\right) \tag{10.12}$$

After two pulses, separated by the time interval T during which the area under consideration has moved a distance $\zeta_1 = VT$, the recording is

$$\left[D(\eta, \zeta) \otimes \text{Rect}\left(\frac{\zeta}{\zeta_0}\right)\right] \otimes [\delta(\zeta) + \delta(\zeta - \zeta_1)] \tag{10.13}$$

After development, the spectrum of the negative is observed, a diaphragm with an aperture to isolate the area under consideration being placed against the negative. The spectrum of the amplitude is

$$\tilde{D}(u, v) \, \text{sinc}\left(\frac{\pi v \zeta_0}{\lambda}\right) [1 + \exp(jkv\zeta_0)] \tag{10.14}$$

and except for a constant factor, the intensity is

$$|\tilde{D}(u, v)|^2 \, \text{sinc}^2\left(\frac{\pi v \zeta_0}{\lambda}\right) \cos^2\left(\frac{\pi v \zeta_1}{\lambda}\right) \tag{10.15}$$

The variations of the intensity are represented in Fig. 158. The spectrum consists of parallel equidistant fringes whose fringe spacing yields the speed V of the area under consideration. The orientation of the fringes yield the direction of the displacement.

If N pulses with duration τ occur during the exposure, and if the pulses are separated by equal intervals of time T, the recording may ve written

$$\left[D(\eta, \zeta) \otimes \text{Rect}\left(\frac{\zeta}{\zeta_0}\right)\right] \otimes \sum_{n=0}^{N} \delta(\zeta - N\zeta_1) \tag{10.16}$$

The intensity of the spectrum is

$$|\tilde{D}(u, v)|^2 \, \text{sinc}^2\left(\frac{\pi v \zeta_0}{\lambda}\right) \left(\frac{\sin(N\pi v \zeta_1/\lambda)}{\sin(\pi v \zeta_1/\lambda)}\right)^2 \tag{10.17}$$

The fringes corresponding to two waves of Fig. 158 are replaced by multiple wave fringes that are finer. In every case, the spectrum is modulated by the

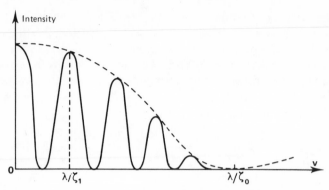

Fig. 158   Fringes in the spectrum of the negative yield the direction and the speed of the object displacement.

function $\mathrm{sinc}^2(\pi v \zeta_0 / \lambda)$, and τ must be small enough with respect to T that the envelope that represents this function (the dotted line in Fig. 158) does not fall off too rapidly. Instead of observing the spectrum as just described, the diaphragm with a small aperture may be placed in the plane of the spectrum instead of against the negative. In this case, the apparatus of Fig. 94 is used and the image of the negative is observed. On the image are seen fringes that represent the component of the velocity in a direction parallel to the azimuth of the aperture. The components of the velocity may be determined in different directions by changing the azimuth of the aperture in the plane of the spectrum.

10.10   An Example of Industrial Application
of Speckle Interferometry

Because of the simplicity of techniques founded on laser speckle, numerous applications are possible. One example is the study of strains in joints of precast reinforced concrete elements. These elements, such as large panels, boards, and supporting walls, are widely used in the construction of buildings. The assembly of these elements, together with the joints that make up the structures, plays an important role in the rigidity of the structures. Displacement measurements made by conventional means do not yield sufficient information about the global behavior of the joints. Speckle interferometry leads to a satisfactory solution.

The principle of the technique is that described in Chapter VII. The experiment is carried out on real joints 2-m long, and not on models. Because of the large surface to be illuminated, a powerful laser must be used. The laser beam goes through anamorphic optics consisting of two cylindrical lenses that spread out the beam only on the surface of the joint. Two exposures are made from the same photographic plate, the first with an unloaded joint and the second with the load added. After development, the negative is observed with the apparatus shown in Fig. 94. For measurement purposes, it is preferable to study the negative point by point as explained in Section 7.1 (Fig. 95). Displacements between 0.04 mm and a few millimeters are easily done. The relative displacements of various parts of the structure under study may thus be determined. The uniformity of the distribution of the load may also be verified.

References

* Adam, F. D. (1972). A study of the parameters associated with employing laser speckle correlation fringes to measure in-plane strain. *A.F.F.D.L.-TR* **72–20.**
* Agarwal, G. S. (1975). Scattering from rough surfaces. *Opt. Commun.* **14,** 161.
* Aggarwal, A. K. (1977). Spatial multiplexing technique to detect sign of in-plane displacement using laser speckle photography. *J. Opt.* **8,** 267.
* Aggarwal, A. K., and Gupta, P. C. (1976). A Fourier transform speckle method to determine the change in angle of illumination. *Opt. Commun.* **17,** 277.
  Archbold, E., and Ennos, A. E. (1972). Displacement measurement from double exposure laser photographs. *Opt. Acta* **19,** 253.
* Archbold, E., and Ennos, A. E. (1974). Applications of holography and speckle photography to the measurement of displacement and strain. *J. Strain Anal.* **9,** 10.
* Archbold, E., and Ennos, A. E. (1975a). Laser photography to measure the deformation of weld cracks under load. *Int. J. Nondestr. Test.* **8,** 181.
* Archbold, E., and Ennos, A. E. (1975b). Two dimensional vibrations analysed by speckle photography. *Opt. Laser Technol.* **7,** 17.
* Archbold, E., Burch, J. M., Ennos, A. E., and Taylor, P. A. (1969a). Visual observations of surface vibration nodal patterns. *Nature (London)* **222,** 263–265.
* Archbold, E., Ennos, A. E., and Taylor, P. A. (1969b). A laser speckle interferometer for the detection of surface movements and vibration. *In* "Optical Instruments and Techniques" (J. H. Dickson, ed.), p. 265. Oriel, Newcastle upon Tyne, England.
  Archbold, E., Burch, J. M., and Ennos, A. E. (1970). Recording of in-plane displacement by double-exposure speckle photography. *Opt. Acta* **17,** 883.
* Arsenault, H. H., and April, G. (1976a). Speckle removal by optical and digital processing. *J. Opt. Soc. Am.* **66,** 177.
* Arsenault, H. H., and April, G. (1976b). Properties of speckle integrated with a fine aperture

* Starred references are not cited in the text.

149

and logarithmically transformed. *J. Opt. Soc. Am.* **66,** 1160.

* Asakura, T., Fuji, H., and Murata, K. (1972). Measurement of spatial coherence using speckle patterns. *Opt. Acta* **19,** 273.

* Barakat, R. (1973). First-order probability densities of laser speckle patterns observed through finite-size scanning aperture. *Opt. Acta* **20,** 729.

* Barakat, R. (1974). Some extreme value statistics of laser speckle patterns. *Opt. Commun.* **10,** 107.

* Beckmann, P., and Spizzichino, A. (1963). "The Scattering of Electromagnetic Waves from Rough Surfaces." Pergamon, Oxford.

* Beddoes, D. R., Dainty, J. C., Morgan, B. L., and Scaddan, R. J. (1976). Speckle interferometry on the 2.5 m Isaac Newton telescope. *J. Opt. Soc. Am.* **66,** 1247.

* Bernard, R., Mas, D., and Valignat, S. (1972). Interférences avec deux images homothétiques superposées. *Opt. Commun.* **5,** 206.

* Breckinridge, J. B. (1976). Interference in astronomical speckle patterns. *J. Opt. Soc. Am.* **6,** 1240.

* Briers, J. D. (1975). A note on the statistics of laser speckle patterns. *Opt. Quantum Electron.* **7,** 422.

  Burch, J. M. (1953). Scatter fringes of equal thickness. *Nature (London)* **171,** 889.

* Burch, J. M. (1962). Scatter fringe interferometry. *J. Opt. Soc. Am.* **52,** 600.

* Burch, J. M. (1971). Laser speckle metrology. *Proc. Soc. Photo-opt. Instrum. Eng. Dev. Holography, Boston, 1971,* pp. 149–156.

* Burch, J. M. (1972). Interferometry with scattered light. *In* "Optical Instruments and Techniques" (J. H. Dickson, ed.), p. 213. Oriel, Newcastle upon Tyne, England.

* Burch, J. M., and Tokarski, J. M. J. (1968). Production of multiple beam fringes from photographic scatters. *Opt. Acta* **15,** 101.

* Burckhardt, C. B. (1970). Laser speckle patterns. A narrow band noise model. *Bell Syst. Tech. J.* **49,** 309–315.

* Butters, J. N. (1971a). Speckle pattern interferometry using video techniques. *Proc. Soc. Photo-opt. Instrum. Eng. 15th Annu. Tech. Symp. Photo-opt. Instrum. for the 70's, Anaheim, 1970,* pp. 77–82.

* Butters, J. N. (1971b). Speckle pattern interferometry using video techniques. *J. Soc. Photo-Opt. Instrum. Eng.* **10,** 5–9.

* Butters, J. N. (1972). Laser holography and speckle patterns in meterological techniques of nondestructive testing. *Int. J. Nondestr. Test.* **4,** 31.

* Butters, J. N., and Leendertz, J. A. (1970). Application of speckle pattern to the simplification of engineering measurement. *S.R.C. Research Report.*

* Butters, J. N., and Leendertz, J. A. (1971). Speckle pattern and holography techniques in engineering metrology. *Opt. Laser Technol.* **3,** 26.

* Butters, J. N., and Leendertz, J. A. (1972a). Application of video techniques and speckle pattern interferometry to engineering measurement. *Proc. Eng. Appl. Holography Symp., Los Angeles, 1972,* pp. 361–375.

* Butters, J. N., and Leendertz, J. A. (1972b). Application of coherent light techniques to engineering measurement. *Appl. Opt.* **11,** 1436.

* Celaya, L., Jonathan, J. M., and Mallick, S. (1976). Velocity contours by speckle photography. *Opt. Commun.* **18,** 496.

* Chakraborty, A. K. (1973). The effect of change of polarisation of the illuminating beam on the microstructure of speckles produced by a random diffuser. *Opt. Commun.* **8,** 366.

  Chandley, P. J., and Welford, W. T. (1975). Optical methods of measuring surface roughness. *Proc. Tenth Congr. Int. Comm. Opt., Prague, 1975,* p. 110.

Chaulnes, duc de (1755). Observations sur quelques experiénces de Newton. *Mém. Anc. Acad. Sci.*, p. 136.

Chiang, F. P., and Juang, R. M. (1976a). Vibration analysis of plate and shell by laser speckle interferometry. *Opt. Acta* **23**, 997.

* Chiang, F. P., and Juang, R. M. (1976b). Laser speckle interferometry for plate bending problems. *Appl. Opt.* **15**, 2199.

* Cloud, G. (1975). Practical speckle interferometry for measuring in-plane deformation. *Appl. Opt.* **14**, 878.

* Colombeau, B., and Froehly, C. (1976). Speckle technique for accurate adjustment of large spacing Fabry–Perot interferometers. *Opt. Commun.* **17**, 284.

Courjon, D., Reggiani, M., and Bulabois, J. (1975). Applications des propriétés des ondes diffusées: détermination de fonctions de transfert optiques et métrologie des vibrations. *Jpn. J. Appl. Phys.* **14**, 311.

* Crane, R. B. (1970). Use of laser-produced speckle pattern to determine surface roughness. *J. Opt. Soc. Am.* **60**, 1658.

* Crosignani, B., Daino, B., and Di Porto, P. (1976). Speckle-pattern visibility of light transmitted through a multimode optical fiber. *J. Opt. Soc. Am.* **66**, 1312.

* Dainty, J. C. (1970). Some statistical properties of random speckle patterns in coherent and partially coherent illumination. *Opt. Acta* **17**, 761.

* Dainty, J. C. (1971). Detection of images immersed in speckle noise. *Opt. Acta* **18**, 327.

* Dainty, J. C. (1972). Coherent addition of a uniform beam to a speckle pattern. *J. Opt. Soc. Am.* **62**, 595.

* Dainty, J. C. (1975). Stellar speckle interferometry. *In* "Laser Speckle and Related Phenomena" (J. C. Dainty, ed.), p. 255. Springer-Verlag, Berlin.

Dainty, J. C. (1976). The statistics of speckle patterns. *Prog. Opt.* **14**, 1.

* Dainty, J. C., and Scaddan, R. J. (1975). Spatial coherence measurements of the atmosphere using a wavefront-folding interferometer. *Proc. Tenth Congr. Int. Comm. Opt., Prague, 1975*, p. 32.

* Dainty, J. C., and Welford, W. T. (1971). Reduction of speckle in image plane hologram reconstruction by moving pupils. *Opt. Commun.* **3**, 289.

* Davenport, W. B., Jr., and Root, W. L. (1958). "An Introduction to the Theory of Random Signals and Noise," p. 182. McGraw-Hill, New York.

Debrus, S., and Grover, C. P. (1971). Correlation of light beams scattered at different angles by a ground glass. *Opt. Commun.* **3**, 340.

* Debrus, S., Françon, M., and Grover, C. P. (1971). Detection of differences between two images. *Opt. Commun.* **4**, 172.

Debrus, S., Françon, M., and Koulev, P. (1974). Extraction de la différence entre deux images. *Nouv. Rev. Opt.* **5**, 153.

* Deitz, P. H. (1975). Image information by means of speckle-pattern processing. *J. Opt. Soc. Am.* **65**, 279.

Duffy, D. E. (1972). Moiré gauging of in-plane displacement using double aperture imaging. *Appl. Opt.* **11**, 1778.

* Dyson, J. (1958). "Concepts of Classical Optics," p. 377. Freeman, San Francisco, California.

Dzialowski, Y., and May, M. (1976a). Correlation of speckle pattern generated by laser point source-illuminated diffusers. *Opt. Commun.* **16**, 334.

* Dzialowski, Y., and May, M. (1976b). Speckle pattern interferometer. *Opt. Commun.* **18**, 321.

* Ek, L., and Molin, N. E. (1971). Detection of the nodal lines and the amplitude of vibration by speckle interferometry. *Opt. Commun.* **2**, 419.

* Elbaum, M., Greenebaum, M., and King, M. (1972). A wavelength diversity technique for reduction of speckle size. *Opt. Commun.* **5**, 171.

Eliasson, B., and Mottier, F. M. (1971). Determination of granular radiance distribution of a diffuser and its use for vibration and analysis. *J. Opt. Soc. Am.* **61**, 559.

* Enloe, L. H. (1967). Noise-like structure in the image of diffusely reflecting objects in coherent illumination. *Bell Syst. Tech. J.* **46**, 1479.

Ennos, A. E. (1975). Speckle interferometry, from laser speckle and related phenomena. *In* "Topics in Applied Physics" (J. C. Dainty, ed.), p. 203. Springer-Verlag, Berlin.

* Erdmann, J. C., and Gellert, R. I. (1976). Speckle field of curved, rotating surfaces of Gaussian roughness illuminated by a laser light spot. *J. Opt. Soc. Am.* **66**, 1194.

Fabry, C., and Perot, A. (1897). Sur les franges des lames argentées et leur application à la mesure des petites épaisseurs d'air. *Ann. Chim. Phys.* **12**, 1.

* Fercher, A. F., and Sprongl, H. (1975). Automatic measurement of vertex refraction with coherent light. *Opt. Acta* **22**, 799.

* Fernelius, M., and Tome, C. (1970). Use of changes in laser speckle for vibration analysis. *J. Appl. Phys.* **41**, 2252.

Forno, C. (1975). White light speckle photography for measuring deformation, strain and shape. *Opt. Laser Technol.* **17**, 217.

* Françon, M. (1973). Optical processing using random diffusers. *Opt. Acta* **20**, 1.

* Françon, M. (1975). Interference and optical processing using random diffusers. *Jpn. J. Appl. Phys.* **14**, 341.

Françon, M., Koulev, P., and May, M. (1976). Speckle photography by birefringent plates. *Opt. Commun.* **17**, 163.

* Fried, D. L. (1976). Statistics of the laser radar cross section of a randomly rough target. *J. Opt. Soc. Am.* **66**, 1150.

* Fried, D. L. (1976). Statistics of Scattering from Moderately Rough Surface. *Asilomar Conf. Speckle Phenom. Opt. Microwaves Acoust., Pacific Grove, California 1976*, p. 24.

* Fujii, H., and Asakura, T. (1974a). Effect of surface roughness on the statistical distribution of image speckle intensity. *Opt. Commun.* **11**, 35.

* Fujii, H., and Asakura, T. (1974b). A contrast variation of image speckle intensity under illumination of partially coherent light. *Opt. Commun.* **12**, 32.

Fujii, H., and Asakura, T. (1975). Statistical properties of image speckle patterns in partially coherent light. *Nouv. Rev. Opt.* **6**, 5.

* Fujii, H., and Asakura, T. (1976). Measurement of surface roughness properties by using image speckle contrast. *J. Opt. Soc. Am.* **66**, 1217.

* Fujii, H., Uozumi, J., and Asakura, T. (1976). Computer simulation study of image speckle patterns with relation to object surface profile. *J. Opt. Soc. Am.* **66**, 1222.

* Fukaya, T., and Tsujiuchi, J. (1975). Characteristics of speckle random pattern and its applications. *Nouv. Rev. Opt.* **6**, 317.

Gabor, D. (1970). Laser speckle and its elimination. *IBM J. Res. Dev.* **14**, 509–519.

* George, N. (1976a). Speckle from rough, moving objects. *J. Opt. Soc. Am.* **66**, 1182.

* George, N. (1976b). Wavelength sensitivity of speckle from an extended object. *Asilomar Conf. Speckle Phenom. Opt. Microwaves Acoust., Pacific Grove, California, 1976*, pp. 24–26.

* George, N., and Christensen, C. R. (1976). Speckle noise in displays. *J. Opt. Soc. Am.* **66**, 1282.

* George, N., and Jain, A. (1972). Speckle in microscopy. *Opt. Commun.* **6**, 253.

* George, N., and Jain, A. (1974). Space and wavelength dependence of speckle intensity. *Appl. Phys.* **4**, 201.

* Goerge, N., and Jain, A. (1975a). Speckle from a cascade of two diffusers. *Opt. Commun.* **15,** 71.
* George, N., and Jain, A. (1975b). Speckle from a thick diffuser. *Proc. Tenth Congr. Int. Comm. Opt., Prague, 1975,* p. 26.
* Gezari, D. Y., Labeyrie, A., and Stachnik, R. V. (1972). Diffraction limited measurements of nine stars with the 200-inch telescope. *Astrophys. J.,* **178,** L71.
* Goedgebuer, U. P., and Vienot, J. C. (1976). Temporal speckle. *Opt. Commun.* **19,** 229.
* Goldfisher, L. I. (1965). Autocorrelation function and power spectral density of laser produced speckle patterns. *J. Opt. Soc. Am.* **55,** 247.
* Goodman, J. W. (1963). Statistical properties of laser speckle patterns. Stanford Electronics Laboratory Tech. Dept. Rep. No. 2301-1.
* Goodman, J. W. (1965). Some effects of target-induced scintillation on optical radar performance. *Proc. IEEE* **53,** 1688.
* Goodman, J. W. (1974). Is surface roughness information obtainable from the power spectral density of a speckle pattern? *Conf. Speckle Phenom., Loughborough Univ. of Technology, Loughborough, England, 1974.*
* Goodman, J. W. (1975a). Probability density function of the sum of N partially correlated speckle patterns. *Opt. Commun.* **13,** 244.
* Goodman, J. W. (1975b). Dependance of image speckle contrast on surface roughness. *Opt. Commun.* **14,** 324.
  Goodman, J. W. (1975c). Statistical properties of laser speckle patterns. *In* "Laser Speckle and Related Phenomena" (J. C. Dainty, ed.), p. 10. Springer-Verlag, Berlin.
  Goodman, J. W. (1976). Some fundamental properties of speckle. *J. Opt. Soc. Am.* **66,** 1145.
* Gough, P. T., and Bates, R. H. T. (1974). Speckle holography. *Opt. Acta* **21,** 243.
  Gregory, D. A. (1976). Basic physical principles of defocused speckle photography. *Opt. Laser Technol.* **8,** 201.
* Grobler, B. (1974). An interesting behavior of speckling in coherent light. *Optik* **40,** 114.
* Groh, G. (1970). Engineering uses of laser produced speckle patterns. *In* "The Engineering Uses of Holography," pp. 483–497. Cambridge Univ. Press, London and New York.
  Grover, C. P. (1972a). A new method of image multiplexing using a random diffuser. *J. Opt. Soc. Am.* **62,** 1071.
* Grover, C. P. (1972b). On the carrier frequency photography using a random diffuser. *Opt. Commun.* **5,** 256.
* Grover, C. P. (1973). Multiple exposure in-line holography using random diffuser. *Appl. Opt.* **12,** 149.
* Hariharan, P. (1975). Speckle-shearing interferometry a simple optical system. *Appl. Opt.* **14,** 2563.
* Hariharan, P., and Hegedus, Z. S. (1974). Reduction of speckle in coherent imaging by spatial frequency sampling. *Opt. Acta* **21,** 345–356.
* Hariharan, P., and Hegedus, Z. S. (1976). Pupil size and speckle statistic for a rough metal surface. *Int. Conf. Appl. Holography Opt. Data Process., Jerusalem, 1976,* p. 319.
* Hariharan, P., Steel, W. H., and Wyant, J. C. (1974). Double grating interferometer with variable lateral shear. *Opt. Commun.* **11,** 317.
  Herschel, J. (1830). Encyclopédie Metropolitana, Vol. 2, Part 2, p. 473. Baldwir and Crodok, Cambridge, England.
* Hung, Y. Y. (1974). A speckle shearing interferometer, a tool for measuring derivative of surface displacements. *Opt. Commun.* **11,** 132.
  Hung, Y. Y., and Taylor, C. E. (1973). Speckle-shearing interferometric camera. A tool for

measurement of derivatives of surface-displacement. *Proc. Soc. Photo-Opt. Instrum. Eng.* **41**, 169.

* Hung, Y. Y., Rowlands, R. E., and Daniel, I. M. (1975). Speckle-shearing interferometric technique: a full-field strain gauge. *Appl. Opt.* **14**, 618.

* Ichioka, Y. Image formation of partially coherent diffuse object. *J. Opt. Soc. Am.* **64**, 919.

* Ichioka, Y., Yamamoto, K., and Suzuki, T. (1975). Effect of the spatial frequency contents of the object of reducing speckle in partially coherent illumination. *Jpn. J. Appl. Phys.* **14**, 317.

* Ingelstam, E., and Ragnarsson, S. (1972). Eye refraction examined by help of speckle pattern produced by coherent light. *Vision Res.* **12**, 411.

* Jain, A. (1976). Radar speckle reduction in synthetic aperture radar processors by a moving diffuser. *Opt. Commun.* **20**, 239.

* Jakeman, E., Whirter, J. G., and Pusey, P. N. (1976a). Enhanced fluctuations in radiation scattered by moving random phase screen. *J. Opt. Soc. Am.* **66**, 1175.

* Jakeman, E., Pike, E. R., Parry, G., and Saleh, B. (1976b). Speckle patterns in polychromatic light. *Opt. Commun.* **19**, 359.

* Jones, R. (1976). The design and application of a speckle pattern interferometer for total plane strain field measurement. *Opt. Laser Technol.* **8**, 215.

* Jones, R., and Butters, J. N. (1975). Some observations on the direct comparison of the geometry of two objects using speckle pattern interferometric contouring. *J. Phys. E* **8**, 231.

Joyeux, D., and Lowenthal, S. (1971). Real time measurement of angström order transverse displacement or vibrations by use of laser speckle. *Opt. Commun.* **4**, 108–112.

* Kallard, T. (1977). "Exploring Laser Light." Optosonic Press, New York.

* Karo, D. P., and Schneiderman, A. M. (1976). Speckle interferometry lens-atmosphere MTF measurements. *J. Opt. Soc. Am.* **66**, 1252.

* Kermish, D. (1974). Speckle reduction in holography by means of random spatial sampling. *Appl. Opt.* **13**, 1000.

* Knox, K. T. (1976). Image retrieval from astronomical speckle patterns. *J. Opt. Soc. Am.* **66**, 1236.

* Komatsu, S., Yamaguchi, I., and Saito, H. (1976). Velocity measurements using structural change of speckle. *Opt. Commun.* **18**, 314.

* Kopf, U. (1972a). Visualization of phase-objects by spatial filtering of laser speckle photographs. *Optik* **36**, 592–595.

Kopf, U. (1972b). Darstellung von phasenobjekten durch Kohärent-optische filterung von laser-granulations-photographien. *Optik* **36**, 592.

Kopf, U. (1974). Application of speckling in carrier-frequency photography. *Int. Opt. Comput. Conf., Zurich, 1974*, p. 862-3C.

* Korff, D. (1963). Analysis of a method for obtaining near-diffraction limited information in the presence of atmospheric turbulence. *J. Opt. Soc. Am.* **63**, 971.

* Korff, D., Dryden, G., and Muller, M. G. (1972). Information retrieval from atmospheric induced speckle patterns. *Opt. Commun.* **5**, 187.

* Kozma, A., and Christensen, C. R. (1976). Effects of speckle on resolution. *J. Opt. Soc. Am.* **66**, 1257.

* Labeyrie, A. (1970). Attainment of diffraction limited resolution in large telescopes by Fourier analysing speckle patterns in star images. *Astron. Astrophys.* **6**, 85.

Labeyrie, A. (1974). Observations intérferométriques au Mont Palomar. *Nouv. Rev. Opt.* **5**, 141.

* Labeyrie, A. (1975). Measurement of stellar angular diameters by speckle interferometry. *ICO Jpn. J. Appl. Phys.* **14,** 283.

Labeyrie, A. (1976). High resolution techniques in optical astronomy. *Prog. Opt.* **14,** 47.

* Lahart, M. J., and Marthay, A. A. (1975). Image speckle patterns of weak diffusers. *J. Opt. Soc. Am.* **65,** 769.

Leendertz, J. A. (1970). Interferometric displacement measurement on scattering surfaces utilizing speckle effect. *J. Phys. E* **3,** 214.

Leendertz, J. A., and Butters, J. N. (1971). A double exposure technique for speckle pattern interferometry. *J. Phys. E* **4,** 277.

Leendertz, J. A., and Butters, J. N. (1973). An image-shearing speckle pattern interferometer for measuring bending moments. *J. Phys. E* **6,** 1107.

Leger, D. (1976). Deux méthodes de mesures de rugosité par corrélation de speckle. Thèse, Université d'Orsay, France. (unpublished).

* Leger, D., and Perrin, J. C. (1976). Real-time measurement of surface roughness by correlation of speckle. *J. Opt. Soc. Am.* **66,** 1210.

Leger, D., Mathieu, E., and Perrin, J. C. (1975). Optical surface roughness determination using speckle correlation technique. *Appl. Opt.* **14,** 872.

* Lewis, R. W. (1973). Redundancy modulation for coherent imaging systems. *Opt. Commun.* **7,** 22–25.

Lifchitz, A., and May, M. (1972). Phénomènes de diffraction à 3 dimensions. *Opt. Acta* **19,** 187.

Liu, C. Y. C., and Lohmann, A. W. (1973). High resolution image formation through the turbulent atmosphere. *Opt. Commun.* **8,** 372.

Lohmann, A. W., and Weilgelt, G. (1975a). The measurement of motion trajectories by photography. *Opt. Commun.* **14,** 252.

* Lohmann, A. W., and Weilgelt, G. (1975b). Large field interferometry. *Proc. Tenth Congr. Int. Comm. Opt., Prague, 1975*, p. 110.

Lohmann, A. W., and Weigelt, G. P. (1976). Speckle methods for the display of motion paths. *J. Opt. Soc. Am.* **66,** 1271.

* Lowenthal, S., and Arsenault, H. H. (1970). Image formation for coherent diffuse objects: statistical properties. *J. Opt. Soc. Am.* **60,** 1478.

* Lowenthal, S., and Joyeux, D. (1971). Speckle removal by a slowly moving diffuser associated with a motionless diffuser. *J. Opt. Soc. Am.* **61,** 847.

* McKechnie, T. S. (1974a). Measurement of some second order statistical properties of speckle. *Optik* **39,** 258.

* McKechnie, T. S. (1974b). Statistics of coherent light speckle produced by stationary and moving apertures. Ph.D. thesis, Dept. of Physics, Imperial College, London. (unpublished).

* McKechnie, T. S. (1974c). Reduction of speckle by a moving aperture: theory and measurement. *Optik* **41,** 34.

* McKechnie, T. S. (1975a). Reduction of speckle in an image by a moving aperture: Second order statistics. *Opt. Commun.* **13,** 29–34.

* McKechnie, T. S. (1975b). Reduction of speckle in an image by a moving aperture: First order statistics. *Opt. Commun.* **13,** 35–39.

Mallick, S., and Roblin, M. L. (1972). Speckle pattern interferometry applied to the study of phase objects. *Opt. Commun.* **6,** 45.

* Marathay, A. S., Heiko, L., and Zuckerman, J. L. (1970). Study of rough surfaces by light scattering. *Appl. Opt.* **9,** 2470.

Marom, E., and Kasher, I. (1977). Optimal distribution of multiple exposures in speckled image subtraction setups. *Nouv. Rev. Opt.* **8**, 1.

Martin, F., Borgnino, J., and Roddier, F. (1975). Localisation de couches turbulentes atmosphériques par traitement optique de clichés d'ombres volantes stellaires. *Nouv. Rev. Opt.* **6**, 15.

* Mas, G., Palpaguer, M., and Roig, J. (1969). Spectre de granularité d'une plage diffusante éclairée en lumière cohérente. Calcul de contraste. *C. R. Acad. Sci.* **269B**, 633.

May, M., and Françon, M. (1976). Correlation and information processing using speckles. *J. Opt. Soc. Am.* **66**, 1275.

Mendez, J. A., and Roblin, M. L. (1974). Relation entre les intensités lumineuses produites par un diffuseur dans deux plans parallèles. *Opt. Commun.* **11**, 245.

Mendez, J. A., and Roblin, M. L. (1975a). Utilisation des franges d'intérference en lumière diffuse pour l'étude de l'état de surface d'un diffuseur, *Opt. Commun.* **13**, 142.

Mendez, J. A., and Roblin, M. L. (1975b). Variation du speckle en presence d'une rotation de l'objet diffusant. *Opt. Commun.* **15**, 226.

* Miller, M. G., and Korff, D. (1964). Resolution of partially coherent objects by use of speckle interferometry. *J. Opt. Soc. Am.* **64**, 155.

* Miller, M. G., Schneiderman, A. M., and Kellen, P. F. (1975). Second-order statistics of laser speckle patterns. *J. Opt. Soc. Am.* **65**, 779.

* Miyake, K. P., and Tamura, H. (1976). Localization of fringes observed through double-exposure speckle photograph. *Int. Conf. Appl. Holography Opt. Data Process., Jerusalem, 1976*, p. 333.

Mohon, W. N., and Rodeman, A. N. (1973). Laser speckle for determining ametropia and accomodation response of the eye. *Appl. Opt.* **12**, 783.

* Myung, H. L., Holmes, J. F., and Kerr, J. R. (1976). Statistics of speckle propagation through the turbulent atmosphere. *J. Opt. Soc. Am.* **66**, 1164.

* Nagata, K., Yoshida, N., and Nishiwaki, J. (1970). Ensemble averaged coherence function of light reflected from rough surface: Determination of its correlation length. *Jpn. J. Appl. Phys.* **9**, 505.

* Nagata, K., Unehara, T., and Nishiwaki, J. (1973). The determination of rms roughness and correlation length of rough surface by measuring spatial coherence function. *Jpn. J. Appl. Phys.* **12**, 694.

Newton, I. (1931). "Optics," 4th ed., p. 289. C. Bell and Sons, London.

* Ogiwara, H., and Ukita, H. (1975). A speckle pattern velocimeter using a periodical differential detector. *Jpn. J. Appl. Phys.* **14**, 307.

* Ohtsubo, J., and Asakura, T. (1975a). Statistical properties of speckle intensity variations in the diffraction field under illumination of coherent light. *Opt. Commun.* **14**, 30.

* Ohtsubo, J., and Asakura, T. (1975b). Statistical properties of speckle intensity variations in the diffraction field under illumination of partially coherent light. *Nouv. Rev. Opt.* **6**, 189.

* Ohtsubo, J., Fujii, H., and Asakura, T. (1975). Surface roughness measurements by using speckle pattern. *Jpn. J. Appl. Phys.* **14**, 293.

* Parry, G. (1974a). Some effects of temporal coherence on the first order statistics of speckle. *Opt. Acta* **21**, 763.

* Parry, G. (1974b). Some effects of surface roughness on the appearance of speckle in poly-chromatic light. *Opt. Commun.* **12**, 75.

* Parry, G. (1975). Speckle patterns in partially coherent light from laser speckle and related phenomena. *In* "Laser Speckle and Related Phenomena" (J. C. Dainty, ed.), p. 77. Springer-Verlag, Berlin.

* Pearson, A., Kokorowski, C. R., and Pedinoff, M. E. (1976). Effects of speckle in adaptative optical systems. *J. Opt. Soc. Am.* **66,** 1261.

* Pearson, J. E. (1976). Effects of speckle in adaptive optical systems. *Asilomar Conf. Speckle Phenom. Opt., Microwaves Acoust., Pacific Grove, California, 1976,* p. 1, 261.

* Pedersen, H. M. (1974). The roughness dependence of partially developed, monochromatic speckle patterns. *Opt. Commun.* **12,** 156.

* Pedersen, H. M. (1975a). On the contrast of polychromatic speckle patterns and its dependance on surface roughness. *Opt. Acta* **22,** 15.

* Pedersen, H. M. (1975b). Second order statistics of light diffracted from Gaussian, rough surfaces with applications to the roughness dependance of speckles. *Opt. Acta* **22,** 523.

* Pedersen, H. M. (1976a). Theory of speckle dependence on surface roughness. *J. Opt. Soc. Am.* **66,** 1204.

* Pedersen, H. M. (1976b). Object roughness dependence of partially developed speckle patterns in coherent light. *Opt. Commun.* **16,** 63.

* Politch, J. Fringe contrast improvement in speckle interferometry of unpolished metallic surfaces. Dept. of Physics, Technion, I.I.T., Technion City, Haifa, Israel.

* Porcello, J. L., Massey, N. G., Innes, R. B., and Marks, J. M. (1976). Speckle reduction in synthetic-aperture radars. *J. Opt. Soc. Am.* **66,** 1305.

Raman, Sir C. V., Palit, M. A., and Datta, G. L. (1921). On Quetelet's rings and other allied phenomena. *Philos. Mag.* **6,** 826.

* Rawson, E. G., Nafarrate, A. B., Norton, R. E., and Goodmann, J. W. (1976). Speckle-free rear-projection screen using two close screens in slow relative motion. *J. Opt. Soc. Am.* **66,** 1290.

* Rigden, J. D., and Gordon, E. I. (1962). The granularity scattered optical maser light. *Proc. I.R.E.* **50,** 2367–2368.

Roblin, M. L., Schalow, G., and Chourabi, R. (1977). Interférométrie différentielle des aberrations d'un systéme optique par photographie de speckles. *Nouv. Rev. Opt.* **8,** 144.

* Roddier, F. (1974). Speckle interferometry through small multiple apertures: Michelson stellar interferometry and apertures synthesis in optics. *Opt. Commun.* **10,** 103.

* Saleh, B. E. A. (1975). Speckle correlation measurement of the velocity of a small rotating rough object. *Appl. Opt.* **14,** 2344.

Sawatari, T., and Elek, A. C. (1973). Image plane detection using speckle patterns. *Appl. Opt.* **12,** 881.

* Schneiderman, A. M., Kellen, P. F., and Miller, M. G. (1975). Laboratory simulated speckle interferometry. *J. Opt. Soc. Am.* **65,** 1287.

Scott, R. M. (1969). Scatter plane interferometry. *Appl. Opt.* **8,** 531.

* Shack, R. V. (1968). Geometric VS diffraction prediction of properties of a star image in the presence of an isotropic random wavefront disturbance. Univ. of Arizona, Tucson, Tech. Rep. No. 32.

Shoemaker, A. H., and Murty, M. V. R. K. (1966). Some further aspects of scatter-fringe interferometry. *Appl. Opt.* **5,** 603.

Sprague, R. A. (1972). Surface roughness measurement using white light speckle. *Appl. Opt.* **11,** 2811.

Stetson, K. A. (1971). New design for laser image-speckle interferometer. *Opt. Laser Technol.* **3,** 220.

* Stetson, K. A. (1974). Analysis of double exposure speckle photography with two-beam illumination. *J. Opt. Soc. Am.* **64,** 857.

* Stetson, K. A. (1975). A review of speckle photography and interferometry. *Opt. Eng.* **14**, 482.

* Stetson, K. A. (1976). Problem of defocusing in speckle photography, its connection on hologram interferometry, and its solutions. *J. Opt. Soc. Am.* **66**, 1267.

Stokes, Sir G. (1851). On the colours of thick plates. *Trans. Cambridge Philos. Soc.* **9**, 147.

* Stone, J. M. (1963). "Radiation and Optics," pp. 146–148. McGraw-Hill, New York.

* Taki, N. (1974). Statistics of dynamic speckles produced by a moving diffuser under the Gaussian beam laser illumination. *Jpn. J. Appl. Phys.* **13**, 2025.

Tanner, L. H. (1969). Three-beam holography with scatter plates. *J. Phys. E* **2**, 288.

* Thinh, V. N., and Tanaka, S. (1976). Measurement of the spectral distribution of a multimode dye laser light by using speckle patterns. *Opt. Commun.* **19**, 378.

* Thinh, V. N., and Tanaka, S. (1977). Measurement of arbitrary parallel displacement of a rigid body using polychromatic speckle pattern. *Opt. Commun.* **20**, 367.

Tiziani, M. J. (1971). Application of speckling for in-plane vibration analysis. *Opt. Acta* **18**, 891.

Tiziani, M. J. (1972a). Analysis of mechanical oscillations by speckling. *Appl. Opt.* **11**, 2911.

* Tiziani, M. J. (1972b). A study of the use of laser speckle to measure small tilts of optically rough surfaces accurately. *Opt. Commun.* **5**, 271.

Tribillon, G. (1974). Corrélation entre deux speckles obtenus avec deux longueurs d'onde. Application à la mesure de la rugosité moyenne. *Opt. Commun.* **11**, 172.

* Tribillon, G., and Garcia, M. (1977). Speckle image à deux longueurs d'onde. *Opt. Commun.* **20**, 229.

* Waterworth, P., and Reid, D. C. (1975). Loose contact detection using laser speckle. *Opt. Laser Technol.* **7**, 135.

* Weigelt, G. P. (1975). Large field speckle interferometry. *Optik* **43**, 111.

* Weigelt, G. P. (1976). Real time measurement of the motion of a rough object by correlation of speckle patterns. *Opt. Commun.* **19**, 223.

* Weigelt, G. P. (1977). Modified astronomical speckle interferometry. *Opt. Commun.* **21**, 55.

* Welford, W. T. (1971). Time-averaged images produced by optical systems with time-varying pupils. *Opt. Commun.* **4**, 275.

* Welford, W. T. (1975). First order statistics of speckle produced by weak scattering media. *Opt. Quantum Electron.* **7**, 413.

* Welford, W. T. (1976). Speckle in images. *J. Opt. Soc. Am.* **66**, 1172.

* Worden, S. P., Lynds, C. R., and Harvey, J. W. (1976). Reconstructed images of alpha Orions using stellar speckle interferometry. *J. Opt. Soc. Am.* **66**, 1243.

* Yamaguchi, I. (1972). Speckling in the diffraction and the image field of rough objects. Derivation of the statistical functions for linear systems. *Optik* **35**, 591.

* Yamaguchi, I., Komatsu, S., and Saito, H. (1975). Dynamics of speckles produced by a moving object and its applications. *Jpn. J. Appl. Phys.* **14**, 301.

Young, T. (1802). On the theory of light and colours. *Philos. Trans. R. Soc. London*, p. 41.

* Yu, F. T. S., and Wang, E. Y. (1973). Speckle reduction in holography by means of random spatial sampling. *Appl. Opt.* **12**, 1656–1659.

* Zelenka, J. S. (1976). Comparison of continuous and discrete mixed-integrator processors. *J. Opt. Soc. Am.* **66**, 1295.

Index

A

Abbe's experiment, 73, *see also* Diffraction
Aberrations, 137
  spherical, 139
Airy disk, 3, 111, *see also* Diffraction
Angle of incidence, 61, 122
Annular aperture, diffraction from, 4
Annular radius, 37
Atmospheric turbulence, 111, 141
Autocorrelation function, 13–14, 63

B

Birefringent plates, 55
Brain, x ray, 81
Burch and Tokarski experiment, 50
Burch interferometer, 41

C

Circle function, 3
Coding, *see* Image

D

Defocusing, 5, 28

Deformations, 89
Delta function, 8
Deviations
  measurements with two wavelengths, 134
  from shape, 121
  from surface, 124
Difference between images, 75
Diffraction
  patterns, 1, 4
  from similar apertures, 22
  in three dimensions, 16
Diffuser, 69, *see also* Object
  auxiliary, 99, 104
  identical, 25, 40
  slopes of, 107
Dirac comb, 54
Displacement
  axial, 24, 31, 47, 96, 99
  continuous, 58
  lateral, 25, 42, 89–95
Double exposure, 50, *see also* Multiple exposures

E

Eye, ametropia, 140

F

Focusing with speckle, 139
Fourier transform, 1, *see also* Spectrum of a
    random distribution
    of a diffuser, 51
    of star speckle, 116
Fraunhofer diffraction, 21
Fresnel diffraction, 20
Fringe
    contrast, 53, 60, 124
    profile improvement, 79
Fringes
    circular, 36–38, 65–72
    elliptical, 72
    hyperbolic, 72
    inside speckle, 43–45
    Young's, 12

G

Grain structure, 9, *see also* Speckle

H

Holograms, Gabor, 70
Hung and Taylor interferometer, 108
Huygens–Fresnel principle, 1

I

Identical patterns, spectrum of, 53
Illumination, two beams, 93
Image
    coding, 83
    modulation, *see* Image processing
    multiplexing, 83–86
    processing, 73–75
Interference, *see* Fringes
Interferometer, *see* specific type
Interferometry, 133
Isophotes, 79

L

Leendertz interferometer, 96

M

Moiré pattern, 95

Multiple exposures, 48, 53
Multiple waves, *see* Fringes

O

Object
    diffuse reflecting, 14
    diffuse transparent, 14, 24, *see also* Diffuser

P

Partial coherence, *see* Sources
Polarized light, 60

R

Reference surface, 95
Rings, *see* Fringes
Rotation, 61
    of aperture, 86
    measurement, 100, 101

S

Scattered light, interference with, 33, *see also*
    Speckle
Secondary maxima, 54
Shadows, flying, 141
Slit
    diffraction from, 4, 13
    oriented, 85
Slopes of objects, 106, 133
Sources
    coherent, 6, 9
    incoherent, 8, 9
    partially coherent, 9, 128, 129
    random distribution of, 8
Speckle
    auxiliary, 31, 99
    contrast, 129, 130
    correlation, 16, 26, 30, 32, 66, 124
    defocusing, 28, 31, 67
    at a finite distance, 20, 24
    in the focal plane, 95, 98
    grain diameter, 24
    image plane, 1, 14, 15
    laterally shifted, *see* Displacement
    modulation with, 74
    observation, 24

oriented, 85
patterns, 9
    multiple, 49, 53
    recording, *see* specific technique
    structure, 28, 82, *see also* Speckle patterns
    superposition, 83
Spectrum of a random distribution, 9–13, *see*
    also specific object
Star diameters, 112
    Labeyrie's method, 115
    with multiple telescopes, 118
Stars, double, 113
Surface
    roughness, 121, 129
    states, 121

T

Trajectory, diffuse object, 142
Transfer function, 136

Translation, *see* Displacement
Transmittance, amplitude, 49
Turbulence, *see* Atmospheric turbulence
Two shifted patterns, spectrum of, 51

V

Velocity measurements, 144
Vibrations, diffusers, 102, 106, 107

W

Wavelength change, 30, 62, 67
Wavelengths, two, 127, 128
White-light recording, 19

Z

Zero shift, circle of, 62